Technology and the frontiers of knowledge

Technology and the Frontiers of Knowledge

The First Frank Nelson Doubleday
Lectures, at The National Museum
of History and Technology, Smithsonian
Institution

1972–73

TECHNOLOGY AND THE FRONTIERS OF KNOWLEDGE

Foreword: Daniel J. Boorstin

SAUL BELLOW
DANIEL BELL
EDMUNDO O'GORMAN
SIR PETER MEDAWAR
ARTHUR C. CLARKE

THE FRANK NELSON DOUBLEDAY LECTURES—1972–73

Doubleday & Company, Inc., Garden City, New York

The Frank Nelson Doubleday Lectures at the Smithsonian's National Museum of History and Technology honor the memory of a great American publisher. The lectures have been made possible by a gift from Doubleday & Company, Inc. on the seventy-fifth anniversary of the firm.

Daniel Bell's lecture "Technology, Nature, and Society" appeared in *The American Scholar*, June 1973.

Sir Peter Medawar's lecture "Technology and Evolution" appeared in the *Smithsonian* magazine, May 1973.

Library of Congress Cataloging in Publication Data
Main entry under title:

Technology and the frontiers of knowledge.

(Frank Nelson Doubleday lecture series for 1972–73)
CONTENTS: Bellow, S. Literature in the age of technology.—Bell, D. Technology, nature, and society.—O'Gorman, E. History, technology, and the pursuit of happiness. [etc.]
1. Technology—Addresses, essays, lectures.
I. Bellow, Saul. II. Series.
T185.T38 301.24'3
ISBN 0-385-09942-8
Library of Congress Catalog Card Number 73–14077

Printed in the United States of America
First Edition

Contents

Foreword

Daniel J. Boorstin

Frank Nelson Doubleday (1862–1934), in whose honor this series of lectures is named, was one of the most remarkable figures in the history of American publishing. It is appropriate that his name be associated with the Smithsonian Institution, for few men did as much as Doubleday to find new forms of organization and enterprise so that publishing might serve the increase and diffusion of knowledge. He accomplished this with a wit and warmth that accumulated an increasing fund of friendship. He was a catalyst among authors in an age fertile of authors, and became a legend in the trade of making and selling books.

Born in Brooklyn in 1862, Doubleday was a self-made man. Before he was fourteen he had bought a small printing press and was turning out (as he later recalled) "all kinds of things, among them the worst visiting cards the world has ever seen, and hat tips to be pasted in straw hats." Without finishing high school, he made his start in publishing in a three-dollar-a-week job as office boy in the firm that became Charles Scribner's Sons. There he performed brilliantly for eighteen years, and then opened his own firm, in 1897. To mark the seventy-fifth anniversary of that occasion, Doubleday & Company initiated and is supporting these Frank Nelson Doubleday Lectures. Frank Nelson Doubleday's accomplishments crossed the whole spectrum of publishing. Before his death, in 1934, virtually every form of the published word (with the exception only of the daily newspaper) had been invigorated by him. He was a publisher of magazines, for example, *World's Work* (a pioneer American journal of social criticism and social science), *American Home* (a prototype of the journal of homemakers), and *Frontier* and *West* (a new sort of fiction-journal exploiting characteristically American themes).

His most potent experiments and innovations were in the world of books. On the death of William Heinemann, in 1920, he became the principal owner of that great English publishing house, and forayed into international publishing. He was adept and imaginative at what he called "The Difficult Art of Selling Books." He developed the retail bookshop into a publisher's laboratory and was among the first to recognize the new possibilities in a potent American institution, the Book Club. He pioneered mail-order and subscription bookselling, which he developed into a fine art. According to one of his authors, Christopher Morley, he pursued the idea "that publishing should be essentially an intelligently conducted commerce, not a form of aesthetic Bohemianism."

One of his most remarkable achievements was his ability, even as his enterprises grew, to preserve an intimacy with his authors, who included some of the great originals of his age. One of them, Rudyard Kipling, affectionately transliterated F.N.D. to *Effendi* (the Turkish word for "Master"). He was one of the first publishers to send authors royalty statements that were substantiated by outside accountants. But there were more sentimental reasons, too, why authors loved him, which included the spontaneity of his affection and his respect for the special talent of each of his authors. These covered the range of thought and literature in his day and included (among others) Joseph Conrad, Edna Ferber, Mary Wilkins Freeman, Hamlin Garland, Henry George, Ellen Glasgow, O Henry, Sinclair Lewis, William McKinley, Frank Norris, Kathleen Norris, John D. Rockefeller, and Theodore Roosevelt.

The title of the continuing Frank Nelson Doubleday Series —"Frontiers of Knowledge"—is appropriate for a great publisher's memorial and a memorial to any publisher who has done his job. And the special concern of the 1972–73 series is Technology and the Frontiers of Knowledge. This is peculiarly appropriate to commemorate the imagination and energy of a man who drew the resources of twentieth-century technology into a wider, more effective diffusion of one of the most venerable vehicles of knowledge: the book.

To help us see into the meanings of modern technology for explorations on the frontiers of knowledge, The National Museum of History and Technology has chosen five thinkers whose concerns touch many of the humanistic and scientific achievements of our age. We have not sought to impose a rigid framework on their reflections. Rather, we have selected them for the individuality of their views, for the acuteness of their insights, and for their ability to interpret to the nonspecialist some of their own special concerns. Saul Bellow, one of the most important novelists of twentieth-century America, seeks the meaning of technology for the literary artist. Daniel Bell, an eminent sociologist, examines the role of technology in forming our knowledge of society. Edmundo O'Gorman, an accomplished historian and archaeologist, suggests how American technology has shaped our ways of viewing the past. Sir Peter Medawar, an adventuring biologist, probes the significance of technology for human evolution. And Arthur C. Clarke, a science writer and skilled imaginer of the future, brings our series to a close with his speculations on technology and the limits of knowledge.

While there is no formal schema to these essays, they are held together, we hope, by a common quest for a better understanding of some of the more troublesome and more promising innovations of modern times. We hope that this book will be a starting point for the reader to do some of his own exploring in the whole mysterious and enticing world which is the province of The National Museum of History and Technology.

Literature in the Age of Technology

Saul Bellow

Nineteenth-century writers disliked or dreaded science and technology. Edgar Allan Poe, discovering that scientific attitudes could be richly combined with fantasy, created science fiction. A Shelley experimented romantically with chemicals; a Balzac thought himself a natural historian or social zoologist; but, for the most part, science, engineering, technology horrified writers. To mechanical energy and industrial enterprise, mass production, they opposed feeling, passion, "true work," artisanship, well-made things. They turned to nature, they specialized in the spirit, they valued love and death more than technical enterprise. Writers then preferred, and still do prefer, the primitive, exotic, and irregular. These romantic attitudes produced masterpieces of literature and painting. They produced also certain cultural platitudes. The platitude of dehumanizing mechanization formed on the one side. Equally platitudinous, the vision of a new age of positive science and of rational miracles, of progress, a progress that made art as obsolete as religion, filled up the other horizon. The platitude of a dehumanized technology gives us, today, novels whose characters are drug-using noble savages, beautiful, mythical primitives who fish in the waters dammed up by mighty nuclear installations. And, as power-minded theoreticians see it, the struggle between old art and new technology has ended in the triumph of technology. The following statement, and it is a typical one, is made by Mr. Arthur C. Clarke in a book called *Report on Planet Three:*

> It has often been suggested that art is a compensation for the deficiencies of the real world; as our knowledge, our power and above all our maturity increase, we will have less and less need for it. If this is true, the ultra-intelligent machine would have no use for it at all.
>
> Even if art turns out to be a dead end, there still remains science. . . .

This statement by a spokesman of the "victorious" party is for several reasons extraordinarily silly. First, it assumes that art belongs to the childhood of mankind, and that science is identical with maturity. Second, it thinks art is born in weakness and fear. Third, in its happy worship of "ultra-intelligent" machines, it expresses a marvelous confidence in the ability of such machines to overcome all the deficiencies of the real world. Such optimistic rationalism is charming, in a way. Put it into rhyming verses and it may sound a lot like Edgar A. Guest. Edgar felt about capitalism and self-reliance precisely as Arthur C. Clarke feels about the supertechnological future. They share a certain expansiveness, the intoxication of the winner, the confidence of the great simplifier. Mr. Clarke says in effect, "Don't worry, dear pals, if art is a dead end, we still have science. Pretty soon we won't need Homer and Shakespeare, Monteverdi and Mozart. Thinking machines will give us all the wisdom and joy we want, in our maturity."

I have chosen a different sort of theorist to put the question from another angle. In the *Atlantic* of July 1972, Mr. Theodore Roszak takes issue with Robert J. Good, a professor of chemical engineering. In a letter to the magazine Professor Good, of the party of science and technology, says that it is sad to see modern intellectuals "cutting off their own roots in rationality." Mr. Roszak tries to deal gently with the professor. He says with a pious tremor,

> What can one do, even in radical dissent, but handle with affectionate care so noble and formidable a tradition within our culture—even knowing it is a tragic error and the death

> of the soul? It is not primarily science I pit myself against in what I write. Rather, the wound I seek to heal is that of psychic alienation: the invidious segregation of humanity from the natural continuum, the divorce of visionary energy from intellect and action. What Professor Good disparages as the irrational (Lord! must I, too, centuries after Blake, repeat the lesson in this limp prose?) is a grand spectrum of human potentialities. When rationality is cut away from that spectrum, then the life of Reason becomes that mad rationality which insists that only what is impersonal and empirical, objective and quantifiable is real—*scientifically* real. Believe that and you are not far from tabulating the tragedies of our existence by way of body counts and megadeaths and chemical imbalances within our neural circuitry.

Of course such issues cannot be discussed without invoking Vietnam, or whatever is the most monstrous topic of the day. For ideologists in all fields the political question is always hugely, repulsively, squatting behind a paper screen.

But when we have cited the argument that ultra-intelligent technology has no need of art, and the counterargument that creativity is needed to heal psychic alienation and keep us from criminal wars, we have not altogether exhausted our alternatives. There is a third alternative, which has nothing much to do with compensation for the deficiencies of the world or with society's health. This alternative holds that man is an artist, and that art is a name for something always done by human beings. The technological present may be inhospitable to this sort of doing, but art can no more be taken from humankind than faces and hands. The giving of weight to the particular, and the tendency to invest the particular with resonant meanings, cannot be driven out by the other tendency: to insist on the finitude of the finite and to divest it of awe and beauty.

Is the film 2001, with a sinister computer that speaks in a homosexual voice, a forerunner of the new maturity? Will this sort of drama replace *Othello*, in which an immature husband murders his wife—I call Othello immature, because I have been told by a famous progressive psychiatrist that in future ages, with sexual jealousy gone, Othello will be incomprehensible. Scientism dearly loves to speak of the childish past, the grave future. When I hear people invoking the maturity of future generations, I think of a conversation in the *Anti-Memoirs* between Malraux and a parish priest who joined the Resistance and later died fighting the Germans. Malraux asked this man,

> "How long have you been hearing confessions?"
> "About fifteen years."
> "What has confession taught you about men?"
> ". . . First of all, people are much more unhappy than one thinks . . . and then . . . the fundamental fact is that *there's no such thing as a grown-up* person. . . ."

Anyway, romantics hold that it is very dangerous to sanity to deny the child within us, while scientism says that technological progress is about to carry us for the first time into the adult stage. On both sides, intellectuals take positions on what art can or cannot add to human happiness. In most discussions the accent falls on health, welfare, progress, or politics—anything but art. About art itself, most intellectuals know and care little.

Malraux begins his *Anti-Memoirs* with the wise chaplain of the Maquis. He then goes immediately into the subject of memoirs and confessions and discusses the "theatrical self-image" in autobiography. How to reduce to a minimum the theatrical side of one's nature is a matter of great concern to him.

> Once, man was sought in the great deeds of great men; then he was sought in the secret actions of individuals (a change encouraged by the fact that great deeds were often violent, and the newspaper has made violence commonplace).

Malraux then concludes,

> . . . the confessions of the most provocative memorialist seem puerile by comparison with the monsters conjured up by psychoanalytic exploration, even to those who contest its conclusions. The analyst's couch reveals far more about the secrets of the human heart, and more startlingly, too. We are less astonished by Stavrogin's confession [in Dostoievsky's *The Possessed*] than by Freud's *Man with the Rats:* genius is its only justification.

The genius, I take it, belongs to Dostoievsky. I am not quite sure what Malraux means, but I think he is saying that what a psychoanalyst learns about the human heart is far deeper and more curious than anything the greatest novelists can reveal. Perhaps he hints even that the madman is more profoundly creative in his rat-imaginings or grotesque fantasies than the writer, who can only compensate us for the ordinariness of his "vile secrets" or "frightful memories" by the power of his mind ("genius is [his] only justification"). But even if we do not push the matter so far, we can legitimately take Malraux, a novelist, to be saying that in the field of facts the writer is "puerile." He cannot compete in that field with the clinical expert.

So novelists once gave "information" to the public. But when people now really want to know something, they turn to the expert. Universities and research institutes produce masses of experts, and governments license them. The dazzle of expertise blinds the unsure, the dependent, and the wretched. It is not the novelist alone who has lost ground. Expertise has made all opinion shaky, and even powerful men are reluctant to trust their own judgment. In totalitarian countries, where facts are suppressed, writers of exceptional courage still tell the truth in the old way. (Why was it not a Soviet expert who told the world about the Gulag Archipelago?) But in the Free World, novelists are peculiarly inhibited.

Artists were great and highly visible monuments in the nineteenth century. The public listened deferentially to its Victor Hugos and its Tolstoys. But Shaw and Wells were the last of these prestigious literary spokesmen. In the postwar period, only Bertrand Russell and Jean-Paul Sartre appeared before the world in this role, and even if these two had been more consistent and sensible they would not have affected the public as their greater predecessors had done. The era of the writer as public sage and as dependable informant has ended.

A single standard has been set for novelists and for experts: the fact standard. The result of this strict accountability has been to narrow the scope of the novel, to make the novelist doubt his own powers and the right of his imagination to range over the entire world. The authority of the imagination has declined. This has had two remarkable results. Earlier in the century, certain writers rejected the older novel with its more modest objectives. The Dickens sort of novel—*Great Expectations*, say—was replaced by the more comprehensive novel, nothing less than an aesthetic project for encompassing the whole world. Books such as Joyce's *Ulysses* and Proust's *Remembrance of Things Past* do not draw the real world so much as they replace it by aesthetic fiat. Without perhaps intending to be one, Joyce was, in effect, an aesthetic dictator. The century needed a book? He provided one. It was a book that made other books unnecessary. It had taken about twenty years to write *Ulysses* and *Finnegans Wake*, and it should take just as long to read them. If you devoted two decades to Joyce, you'd have no time for other writers. Anyway, your need for literature would be fully satisfied. This was one result of the weakening of the authority of writers and of the power of the literary imagination to command attention—overassertiveness. A more recent result has been surrender. Writers have capitulated to fact, to events and reportage, to politics and demagogy. Modern art has tried to create power for itself on arbitrary terms and has also pursued and worshiped power in its public forms.

Until recent times, the artist's dream sphere was distinctly separate from the practical or mechanical realm of the technician. But in the twentieth century, as Paul Valéry recognized, a change occurred. He wrote in an essay:

> The fabulous is an article of trade. The manufacture of machines to work miracles provides a living to thousands of people. But the artist has had no share in producing these wonders. They are the work of science and capital. The bourgeois has invested his money in phantoms and is speculating on the downfall of common sense.

Yes, technology is the product of science and capital, and of specialization and the division of labor. It is a triumph of the accurate power of innumerable brains and wills acting in unison to produce a machine or a commodity. These many wills constitute a fictive superself astonishingly effective in converting dreams into machines. Literature, by contrast, is produced by the single individual, concerns itself with individuals, and is read by separate persons. And the single individual, the unit of vital being, of nerve and brain, who judges or knows, is happy or mourns, actually lives and actually dies, is unfavorably compared with that fictive superself which, acting in unison and according to plan, produces jet planes, atomic reactors, computers, rockets, and other modern technological wonders. Glamorous, victorious technology is sometimes considered to have discredited all former ideas of the single self.

To theorists of the new, a thing is genuine only if it manifests the new. Valéry, in the essay *Remarks on Progress*, from which I have just quoted, illustrates this attitude remarkably well. He says,

> Men are doubtless developing the habit of considering all knowledge as transitional and every stage of their industry and their relations as provisional. This is new.

And again,

> Suppose that the enormous transformation which we are living through and which is changing us, continues to develop, finally altering whatever customs are left and making a very different adaptation of our needs to our means; the new

> era will soon produce men who are no longer attached to the past by any habit of mind. For them history will be nothing but strange, incomprehensible tales; there will be nothing in their time that was ever seen before—nothing from the past will survive into their present. Everything in man that is not purely physiological will be altered, for our ambitions, our politics, our wars, our manners, our arts are now in a phase of a quick change; they depend more and more on the positive sciences and hence less and less on what used to be. *New facts* tend to take on the importance that once belonged to tradition and *historical facts.*

This is the quintessence of the tradition of the new. By attaching itself to technology, "newness" achieves a result longed for by those thinkers of the previous century who were oppressed by historical consciousness. Karl Marx felt in history the tradition of all the dead generations weighing like a nightmare on the brain of the living. Nietzsche speaks movingly of the tyranny of "it was," and Joyce's Stephen Dedalus also defines history as a nightmare from which we are trying to awaken. The vision of freedom without conditions, a state of perfect and lucid consciousness into which we are released by technological magic from all inertias, is a sort of romance, really, a French intellectual's paradise. But Valéry does not neglect the painful side of this vision:

> . . . one of the surest and cruelest effects of progress is to add a further pain to death, a pain increasing of itself as the revolution in customs and ideas becomes more marked and rapid. It is not enough to perish; one has to become unintelligible, almost ridiculous; and even a Racine or a Bossuet must take his place alongside those bizarre figures, striped and tattooed, exposed to passing smiles, and somewhat frightening, standing in rows in the galleries and gradually blending with the stuffed specimens of the animal kingdom. . . .

So, at the height of technological achievement there blazes the menace of obsolescence. The museum, worse than the grave because it humiliates us by making us dodos, waits in judgment on our ambitions and vanities. Of course, no one wants to suffer the double doom of obsolescence—to be dead and also to be a fossil. Everyone wants to be the friend and colleague of history. And consciously or not, intellectuals try hard to be what Hegel called Historical Men or World-Historical Individuals, those persons through whom truth operates and who have an insight into the requirement of the time, who divine what is ripe for development, the nascent principle, the next necessary thing. They may denounce the nightmare past, but they have also an immortal craving to be in the line of succession, and to prove themselves to be historically necessary. It is these people, lovers of the new, who derive from technological progress a special contempt for the obsolete. The enemies of pastness, even though they tell us that we will depend more and more on the positive sciences and hence less and less on what used to be, insofar as they seek the next necessary development, make their own kind of historical judgment. Intellectuals, when they sense the cruel effects of technological progress, try not only to escape oblivion themselves by association with the next necessary thing, but also to impose oblivion on others—on those writers who fail to recognize that the human condition has been, or will be, completely transformed by science and the revolution of customs, and who, in the old-fashioned solitude of old-fashioned rooms, continue to consider the destinies of old-fashioned individuals and follow their old-fashioned trade (a home industry of the seventeenth century), unaware that the World Spirit has abandoned them.

Unlike Huxley's *Brave New World* or George Orwell's *Nineteen Eighty-four*, Joyce's *Ulysses* is not directly concerned with technology. It remains nevertheless the twentieth century's most modern novel—it is *the* account of human life in an age of artifacts. Things in *Ulysses* are not nature's things. Here the material world is wholly man's world, and all its objects are human inventions. It is made in the image of the conscious mind. Nature governs physiologically, and of course the unconscious remains nature's stronghold, but the external world is a world of ideas made concrete. Between these two powers, nature within, artifacts without, the life of Mr. Leopold Bloom is comically divided. The time is 1904. No one in Dublin has seen Mr. Arthur C. Clarke's ultra-intelligent machine even in a dream, but the age of technology has begun and *Ulysses* is literature's outstanding response to it.

Now what is *Ulysses?* In *Ulysses* two men, Dedalus and Bloom, wander about the city of Dublin on a June day. Mrs. Bloom, a singer, lies in bed, reading, misbehaving, musing and remembering. But nothing that can be thought or said about human beings is left out of this account of two pedestrians and an adulteress. No zoologist could be more explicit or complete than Joyce. Mr. Bloom first thing in the morning, brews the tea, gives milk to the cat, goes to the pork butcher to get meat for his breakfast, carries a tray up to his wife, eats a slightly scorched kidney, goes out to the privy with his newspaper, relieves himself while reading a prize-winning story, wipes his bottom with a piece of the same paper, and then goes out to the funeral of Paddy Dignam. Matters could not be more real.

Now realism in literature is a convention, and this convention postulates that human beings are not what everyone for long centuries conceived them to be. They are something different, and they live in a disenchanted world that exists for no particular purpose that science can show. Still, people continue to try to lead a human life. And this is rather quaint, because man is not the comparatively distinguished creature he thinks himself to be. The commonness of common life was a great burden to nineteenth-century writers. The best of them tried to salvage something from the new set of appalling facts. Coleridge tells us how Wordsworth intended to redeem the everyday by purging the mind of "the lethargy of custom" and showing us the beauty and power of what we call commonplace. He did not do this to the satisfaction of his successors.

English writers of the second half of the century were much more impressed by the weight given to the evil component of the commonplace by their French contemporaries. A novelist like Flaubert saw nothing in the banal average that did not disgust him. But art, virtuosity, language, the famous objectivity, these, after painful struggle, would make commonplace reality yield gold.

Joyce, a Flaubertian to begin with, gives in *Ulysses* the novel's fullest account of human life—within this realistic convention. As he sees it, the material world is now entirely human. Everything about us—clothing, beds, tableware, streets, privies, newspapers, language, thought—is man-made. All artifacts originate in thought. They are thoughts practically extended into matter.

Joyce is the complete naturalist, the artist-zoologist, the poet-ethnographer. His account of Bloom's life includes everything. Everything seems to demand inclusion. No trivialities or absurdities are omitted. Old bourgeois reticences are overrun zigzag. For what, after all, is the important information? No one knows. Anything at all may be important. Freud taught, in *The Psychopathology of Everyday Life*, that the unconscious did not distinguish between major and minor matters, as conscious judgment did, and that the junk of the psyche had the deepest things to tell us. Joyce is the greatest psychic junkman of our age, after Freud. For the last of the facts may be the first. Thus we know the lining of Bloom's hat, and the contents of his pockets; one knows his genitals and his guts, and we are thoroughly familiar with Molly, too, how she feels before her period, and how she smells. And with so much knowledge, we are close to chaos. For what are we to do with such a burden of information? *Ulysses* is a comedy of information. Leopold Bloom lies submerged in an ocean of random facts: textbook tags, items of news, bits of history, slogans, clichés, ditties, operatic arias, saws and jokes, scraps of popular science, and a great volume of superstitions, fantasies, technical accounts of the Dublin water supply, observations about hanged men, recollections of copulating dogs.

The debris of learning, epic, faith, and enlightenment pour over him. In this circumambient ocean, he seems at times to dwell like a coelenterate or a sponge. The man-made world begins, like the physical world, to suggest infinity. The mind is endangered by the multitude of accounts that it can give of all matters. It is threatened with inanity or disintegration.

William James believed that not even the toughest of tough minds could bear to know everything that happened in a single city on a given day. Not everything. No one could endure it. It is probably one of the functions of the nervous system to screen us and to preserve us from disintegrating in the sea of facts. We ourselves, however, seek out this danger, the Faustian dream of omniscience lives on. To Joyce this Faustian omniscience is a deliciously funny theme. —I assume that Joyce knew Flaubert's last novel, of two elderly cranks, Bouvard and Pécuchet, who in retirement try to investigate every branch of knowledge.

At all events, Bloom's mind is assailed and drowned by facts. He appears to acknowledge a sort of responsibility to these facts, and he goes about Dublin doing his facts. This suggests that our scientific, industrial, technical, urban world has a life of its own and that it borrows our minds and souls for its own purposes. In this sense, civilization lives upon Bloom. His mind is overcome by its data. He is the bearer, the servant, the slave of involuntary or random cognitions. But he is also the poet of distractions. If Bloom were only the *homme moyen sensuel*, or everyman, nothing but the sort of person realism describes as "ordinary," he would not be the Bloom we adore. The truth is that Bloom is a wit, a comedian. In the depths of his passivity, Bloom resists. He is said in Dublin to be "something of an artist." To be an artist in the ocean of modern information is certainly no blessing. The artist has less power to resist the facts than other men. He is obliged to note the particulars. One may even say that he is condemned to see them. In the cemetery, Bloom can't help seeing the gravedigger's spade and noting that it is "blueglancing." He is therefore receptively, artistically, painfully immersed in his mental ocean. The fact that he is "something of an artist" aggravates the problems of information. He seeks relief in digression, in evasion, and in wit.

Why is the diversity of data so dazzling and powerful in *Ulysses?* The data are potent because the story itself is negligible. *Ulysses,* as Gertrude Stein once said, is not a "what-happens-next?" sort of book. A "what-happens-next?" story would, like a nervous system, screen out distractions and maintain order.

It is the absence of a story that makes Bloom what he is. By injecting him with purpose, a story would put the world in order and concentrate his mind. But perhaps Bloom's mind is better not ordered. Why should he, the son of a suicide, the father who mourns a dead child, a cuckold, and a Jew in Catholic Dublin, desire moral and intellectual clarity? If his mind were clear, he would be another man entirely. No, the plan of Bloom's life is to be planless. He palpitates among the phenomena and moves vaguely toward resolution. Oh, he gets there, but *there* is a region, not a point. At one of the low hours of his day he thinks, "Nothing is anything." He feels his servitude to the conditions of being. When there is no story, those conditions have it all their own way and one is delivered to despair. The artifact civilization, Joyce seems to tell us, atrophies the will. The stream of consciousness flows full and wide through the willless. The romantic heroes of powerful will, the Rastignacs and the Raskolnikovs, are gone. The truth of the present day is in the little Blooms, whose wills offer no hindrance to the stream of consciousness. And this stream has no stories. It has themes. Bloom does not, however, disintegrate in the thematic flow. Total examination of a single human being discloses a most extraordinary entity, a comic subject, a Bloom. Through him we begin to see that anything can be something.

But the burden of being a Bloom is nevertheless frightful. It is not clear exactly how Joyce would like us to see the Bloom problem. Long passages of *Ulysses* are bound together by slurs (in the musical sense of the word) of ambiguous laughter. It does, however, appear that Joyce expected the individual who has gone beyond the fictions and postures of "individuality" (romantic will, etc.) to be sustained by suprapersonal powers of myth. Myth, rising from the unconscious, is superior to mere "story," but myth will not come near while ordinary, trivial ideas of self remain. The powers of myth can be raised up only when the discredited pretensions of selfhood are surrendered. Therefore consciousness must abase itself, and every hidden thing must be exhumed. Hence Bloom's moments in the privy, his corpse fantasies at Paddy's funeral, his ejaculation as he watches crippled Gerty, his masochistic hallucinations in Nighttown. The old dignities must take a terrific beating in this new version of "the first shall be last and the last first."

How is the power of the modern age to be answered? By an equal and opposite power within us, tapped and interpreted by a man of genius. Joyce performs the part of the modern genius to perfection. This sort of genius, as I see it, comes from the mass without external advantages. Everything he needs is within him. By interpreting his own dreams, he creates a scientific system. By musical spells extracted from his own personality, he hypnotizes the world with his Siegfrieds and Wotans. He is, as Collingwood calls him, the "mystagogue leading his audience . . . along the dark and difficult places of his own mind . . . the great man who (as Hegel says) imposes upon the world the task of understanding him." This task of understanding has certainly been imposed by Joyce. These men of genius take you in their embrace and propose to be everything to you. They are your *Heimat*, your church. You need no other music than theirs, no other ideas, no other analysis of dreams, no other manna. They are indeed stirring and charming. Their charms are hard to get away from. Theirs are the voices under which other voices sink. Once initiated into their mysteries, we do not easily free ourselves.

It is now seventy years since Bloom walked the streets of Dublin. In those seventy years the noise of life has increased a thousandfold. And already at the beginning of the nineteenth century Wordsworth was alarmed by the increase of distractions. The world was too much with us. The clatter of machinery, business, the roar of revolution would damage the inmost part of the mind and make poetry impossible. A century after Wordsworth, ingenious Joyce proposed to convert the threat itself into literature.

What you feel when you read *Ulysses* today is the extent to which a modern society imposes itself upon everyone. The common man, who, in the past, knew little about the great world, now stands in the middle of it. At least he thinks he does, as reader, hearer, citizen, voter, and judge of all public questions. His imagination has been formed to make him think himself in the center. The all-important story appears to belong to society itself. Real interest is monopolized by collective achievements and public events, by the fate of mankind, by a kind of "politics." The voluminous Sunday *Times* is put into our hands together with *Time* and *Newsweek,* while images from television flash before us. This is the week's record of everything of substance relating to the human species. It is about us, our hope for survival, our common destiny. Is it, now? Does this really speak to my condition? Is this mankind, is it *me,* heart, soul, and destiny? The nominally central individual studying the record does not in fact feel central. On the contrary, he feels peculiarly contentless in his public aspect, lacking in substance and without a proper story. A proper story would express his intuition that his own existence is peculiarly significant. The sense that his existence is significant haunts him. He can prove nothing. But the business of art is with this sense, precisely.

Though the sun shines sweetly, the modern mind knows that there are devilish processes of nuclear fusion and staggering explosions in the heavens. So, as mild Bloom goes down the streets, we are aware of a formidable intellect that follows him as he buys a cake of soap. The modernists are learned intellectuals—Viconians like Joyce, Freudians, Marxians, Bergsonians, et cetera. A technological society produces mental artists and an intensely intellectual literary culture. "Most modern masterpieces are critical masterpieces," writes Harold Rosenberg, who thinks that this is as it should be. "Joyce's writing is a criticism of literature, Pound's poetry a criticism of poetry, Picasso's painting a criticism of painting. Modern art also criticizes the existing culture. . . . One keeps hoping that the decline in excellence of people and things is an effect of transition. All we have on the positive side is the individual's capacity for resistance. Resistance and criticism." To hope that the decline in the excellence of people and things (the last an effect of technology) is an effect of transition, shows Mr. Rosenberg's heart to be in the right place. But the emphasis on criticism shows something else, namely a claim for intellectual priority. Art is something that must satisfy the requirements of intellectuals. It must interest intellectuals by being, in the right sense, critical of the existing culture. The fact is that modern art has tried very hard to please its intellectual judges. Intellectual judgment in the twentieth century very much resembles aristocratic taste in the eighteenth—in the sense, only, that artists in both centuries acknowledge its importance. Art in the twentieth century is more greatly appreciated if it is directly translatable into intellectual interests, if it stimulates ideas, if it lends itself to discourse. Because intellectuals do not like to suspend themselves in works of the imagination. They want to talk. Thus they make theology and philosophy out of literature. They make psychological theory. They make politics. Art is one of the principal supports of this social class.

Gide's *The Counterfeiters* is a cultural product as well as a novel; *The Red and the Black* is no such thing. *The Magic Mountain* belongs to intellectual history, a category that does not exist for that excellent book *Little Dorrit*. It never occurred to Dickens to run over into cultural criticism or to be Carlyle and Mill as well as Charles Dickens. But, in the twentieth century, writers are often educated men as well as creators, and in some the education prevails over the creation. There are reasons for this. A burden of "understanding" has been laid upon us by this revolutionary century. What I am trying to indicate is that cultural style is not to be confused with genuine understanding. At the moment, such understanding has few representatives, while cultural style seems to have hundreds of thousands.

One of the problems of literature in this age of technology is the problem of those who preside over literary problems, of specialists, scholars, historians, and teachers. Modern writers, themselves more "intellectual" than those of a century ago, face a public formed, educated, and dominated by professors, "humanists," "anti-humanists," by psychologists and psychotherapists, by the professional custodians of culture, and by ideologists and shapers of the future. This critical public has a thousand *important* (i.e., political and social) questions to answer. It is irritably fastidious. It asks, "To whom should we listen? Who, if anyone, can be read? What is, or can be, really interesting to a modern cultivated intelligence?"

Such questions can only be answered with sadness and sighs. To prove that I am not exaggerating, I shall quote briefly from an essay by Lionel Trilling, in *Commentary*, September 1971, "Authenticity and the Modern Unconscious." In this essay Professor Trilling observes that in this day and age, things being what they are, novels can no longer be Authentic or appeal to Authentic readers. "It is the exceptional novelist today who would say to himself, as Henry James did, that he 'loved the story as story,' by which James meant the story apart from any ideational intention it might have, simply as, like any primitive tale, it brings into play what James called 'the blessed faculty of wonder' Already in James's day, narration as a means by which the reader was held spellbound, as the old phrase put it, had come under suspicion. And the dubiety grew to the point where Walter Benjamin could say some three decades ago that the art of storytelling was moribund."

Here one cries out, "Wait! Who is this Benjamin! Why

does it matter what *he* said?" But intellectuals do refer to one another to strengthen their arguments. It turns out that the late Walter Benjamin objected to storytellers because they had an orientation toward practical interests. Stories, Trilling quotes Benjamin as saying, are likely to contain "something useful." Here what I have called "cultural style" begins to show itself. Modern literary culture, which prides itself on being radical, dissenting, free, has its own orthodoxy. Don't we know how it views the Bourgeois, the Child, the Family, Technology, the Artist, the Useful? We do indeed. The idea of usefulness, Baudelaire said, nauseated him. And there is the foundation of your orthodoxy. Storytellers, Benjamin objects, have "counsel to give" and this giving of counsel has "an old-fashioned ring." Professor Trilling then continues, It is "inauthentic for our time—there is something inauthentic for our time in being held spellbound, momentarily forgetful of oneself, concerned with the fate of a person who is not oneself. . . . By what right, we are now inclined to ask, does the narrator exercise authority over that other person, let alone over the reader, and arrange the confusion between the two, and presume to have counsel to give?"

If there is something old-fashioned and inauthentic for our time in being held spellbound, then Homer and Dostoievsky, whose works hold us spellbound, are inauthentic. The point seems to be that the modern condition is killing certain human activities (arts) once highly valued. For an Authentic modern man, living in a modern technological society, naive self-surrender is impossible. Apparently the question is partly one of authority. "By what right does the narrator" presume to invade our minds, deliberately confuse us, and give counsel? I don't see what good it does to make a political question of this. By what right do our parents conceive us, or we our children? By what right does society teach us a language or give us a culture? If authentic man had no words, he would be unable to express his longing to be so virginal.

But this, I realize, is not quite fair. Professor Trilling wishes to leave the surface of life with its stories and descend into the depths in search of truth and maturity, becoming one of Aristotle's great-souled men. Thus Professor Trilling seems to agree, with Malraux's priest in the Resistance, that there is no such thing as a grown-up person. His position is also close to that of Mr. Arthur C. Clarke, who suggests that art is a compensation for the deficiencies of the real world and that "as our knowledge, our power and above all our *maturity* increase, we will have less and less need for it. . . . the ultra-intelligent machine would have no use for it at all."

So Professor Trilling, moving toward "scientific truth," reports that we can no longer be held spellbound, and Mr. Clarke tells us that we can be redeemed by technology from the childish need for art. Perhaps a modest, fair statement of the case is that human beings have always told stories to one another. By what *right* have they done this, and on what authority? Well, on none, really. They have obeyed the impulse to tell and the desire to hear. Science and technology are not likely to remove this narrative and spellbinding oddity from the soul. The present age has a certain rationalizing restlessness or cognitive irritability; a participatory delirium that makes the arresting powers of any work of art intolerable. The desire to read is itself spoiled by "cultural interests" and by a frantic desire to associate everything with something else and to convert works of art into subjects of discourse. Technology has weakened certain points of rest. Wedding guests and ancient mariners both are deafened by the terrific blaring of the technological band.

In a charming and strange book, the prerevolutionary Russian writer V. V. Rozanov argues against repressive puritanism in words that can be applied more widely and are relevant to the subject of this paper. He writes:

> A million years passed before my soul was let out into the world to enjoy it; and how can I say to her, "Don't forget yourself, darling, but enjoy yourself in a responsible fashion."
>
> No, I say to her: "Enjoy yourself, darling, have a good time, my lovely one, enjoy yourself my precious, enjoy

> yourself in any way you please. And toward evening you will go to God."
> For my life is my day, and it is my day, and not Socrates' or Spinoza's.

Thus to the queen, or tramp who is his soul Rozanov speaks with an erotic-religious aim of some sort. But we can adapt this to our own purpose, saying, "A million years passed before my soul was let out into the technological world. That world was filled with ultra-intelligent machines, but the soul, after all, was a soul, and it had waited a million years for its turn and did not intend to be cheated of its birthright by a lot of mere gimmicks. It had come from the far reaches of the universe and it was interested but not overawed."

Technology, Nature, and Society

Daniel Bell

The Smithsonian Legacy

The terms of the will of James Smithson, as you know, bequeathed the whole of his property to the United States of America, "to found at Washington, under the name of the Smithsonian Institution, an Establishment for the increase and diffusion of knowledge among men." Though the bequest, in one sense, was clear, the effort to implement it led for several decades to many confusions and debates. What is knowledge, and how does one increase it or diffuse it? Some individuals wanted to create a national university, others a museum, still others a library, and others still a national laboratory, an agricultural experiment station, or, with John Quincy Adams, a national observatory. Today we have all these except a national university—though some local patriots might consider my home on the Charles such an institution. And certainly, under Mr. Ripley, the Smithsonian has become "an Establishment."

But if in later years buildings were built and institutions established, the more vexing question, of what knowledge should be increased and promoted, which bedeviled the regents of the Institution, still remains. In the mid-nineteenth century the "promotion of abstract science," as Joseph Henry, the first head of the Institution, put it, dominated the activities of the Smithsonian. But Mr. Henry soon found himself under attack from all sides. There were those like Alexander Dallas Bache, who said that ". . . a promiscuous assembly of those who call themselves men of science would only end in disgrace." Under the new conditions of scientific specialization, he declared, the universal savant was obsolete; the differentiation of scientists from amateurs demanded the material support only of professional research scientists. On the other hand, Horace Greeley, in the New York *Tribune*, accused Mr. Henry of converting the Smithsonian into "a lying-in hospital for a little knot of scientific valetudinarians." The question of what kind of science, theoretical or applied, continues to be refought.

A different, equally familiar issue was the one between men of science and men of letters. Ethics and philosophy, said Rufus Choate of Massachusetts, were as vital as soil chemistry and a knowledge of noxious weeds, and in the debate in the House of Representatives Choate's protégé, Charles W. Upham, representing the men of letters, declared: ". . . vindicate art, taste, learning, genius, mind, history, ethnology, morals—against sciologists, chemists & catchers of extinct skunks."[1]

In the unhappy further differentiation of the world since then, I appear here neither as a man of science nor as a man of letters. Sociologists (the bearers of superficial learning) have become crossed with logomachs (those who contend wearily about words) to create sociologists, that hybrid with a Latin foreword and a Greek root, symbolizing the third culture which has diffused so prodigiously throughout the modern world.

Yet as an intellectual hybrid my provenance may not be amiss. For my theme this evening is the redesign of the intellectual cosmos, the hybrid paths it has taken, and the

[1] My discussion of the Smithsonian legacy and its vicissitudes is taken from A. Hunter Dupree, *Science in the Federal Government* (Harvard University Press, Cambridge, Mass., 1957), Chapter IV, and Howard S. Miller, "Science and Private Agencies," in Van Tassel and Hall (eds.), *Science and Society in the United States* (Dorsey Press, Homewood, Ill., 1960), pp. 195–201.

necessary and hybrid forms it may take. With Mr. Upham's charge in mind, I am prepared to vindicate all his categories, except extinct skunks.

I

The Confusion of Realms

If we ask what uniquely marks off the contemporary world from the past, it is the power to transform nature. We define our time by technology. And until recently we have taken material power as the singular measure of the advance of civilization.

The philosophical justification of this view was laid down a hundred or more years ago by Marx. Man has needs which can only be satisfied by transforming nature, but in transforming nature he transforms himself: as man's powers expand he gains a new consciousness and new needs—technological, psychological, and spiritual—which serve, further, to stimulate man's activity and the search for new powers. Man, thus, is defined not by nature but by history. And history is the record of the successive plateaus of man's powers.[2]

But if it is, as Marx states in *Capital*, that in changing his external environment man changes his own nature, then human nature in ancient Greece must have been significantly different from human nature under modern capitalism, where needs, wants, and powers are so largely different. And if this is so, how is it possible, as Sidney Hook asks, to understand past historical experience in the same way we understand our own, since understanding presupposes an invariant

[2] "Human history may be viewed as a process in which new needs are created as a result of material changes instituted to fulfill the old. According to Marx . . . the changes in the character and quality of human needs, including the means of gratifying them, is the keystone not merely to historical change but to the changes of human nature." Sidney Hook, *From Hegel to Marx* (University of Michigan Press, Ann Arbor, 1962), p. 277.

pattern? This is a problem which confronts not only historical materialism but all philosophies of history.[3]

Marx only once, to my knowledge, in a fragment written in 1857, sought to wrestle with this conundrum; and his answer is extraordinarily revealing:

> It is a well-known fact that Greek mythology was not only the arsenal of Greek art but also the very ground from which it had sprung. Is the view of nature and social relations which shaped Greek imagination and Greek [art] possible in the age of automatic machinery, and railways and locomotives, and electric telegraphs? Where does Vulcan come in as against Roberts & Co.; Jupiter as against the lightning rod; and Hermes as against the Crédit Mobilier? All mythology masters and dominates and shapes the forces of nature in and through imagination; hence it disappears as soon as man gains mastery over the forces of nature. . . . Is Achilles possible side by side with powder and lead? Or is the Iliad at all compatible with printing press and steam press? Do not singing and reciting and the muses necessarily go out of existence with the appearance of the printer's bar, and do they not, therefore, disappear with the prerequisites of epic poetry?
>
> But the difficulty is not in grasping the idea that Greek art and epos are bound up with certain forms of social development. It rather lies in understanding why they still constitute with us a source of aesthetic enjoyment and in certain respects prevail as the standard and model beyond attainment.

The reason, Marx declares, is that such art is the *childhood* of the human race and carries with it all the charm, artlessness, and precocity of childhood, whose truths we sometimes seek to recapture and reproduce "on a higher plane." Why should "the social childhood of mankind, where it had obtained its most beautiful development, not exert an eternal charm as an age that will never return?"[4] That is why we appreciate the Greek spirit.

[3] Sidney Hook, "Materialism," *Encyclopedia of the Social Sciences*, Vol. X (Macmillan, New York, 1933), p. 219.

[4] "Introduction to the Critique of Political Economy." The essay, much of it in the form of notes, was intended as an introduction to

The answer is a lovely conceit. Yet one must know the sources of the argument to understand the consequences. For Marx, this view derived, in the first instance, from the conception of man as *homo faber*, the tool-making animal; the progressive expansion of man's ability to make tools is, therefore, an index of man's powers. A second source of this view was Hegel, who divided history into epochs or ages, each a structurally interrelated whole and each defined by a unique spirit qualitatively different from each other. From Hegel, this view passed over into cultural history, with its periodization of the Greek, Roman, and Christian worlds, and Renaissance, Baroque, Rococo, and Modern styles. Sociologically, Hegel's idea is the basis of the Marxist view of history as successive slave, feudal, bourgeois, and socialist societies. Behind it all is a determinist idea of progress in human affairs, or a *marche générale* of human history, in which rationality in the Hegelian view, or the powers of production in the Marxist conception, are the immanent, driving forces of history that are obedient to a teleology in which anthropology, or a man-centered world, replaces theology, or a God-created world.

Today we know that, of the two views, that of *homo faber* is inadequate and that of the march of society and history is wrong. Man is not only *homo faber* but *homo pictor*, the symbol-producing creature, whose depictions of the world are not outmoded in linear history but persist and coexist in all their variety and multiplicity through the past and present, outside of "progressive" time. As for the nineteenth-century view of society, just as the mechanistic world view of nature has been destroyed by quantum physics, so the determinist

---

the main work of Marx. The posthumous essay was first published by Karl Kautsky, Marx's literary executor, in *Neue Zeit*, the theoretical organ of the German Social Democratic Party, and published in English as an appendix to *A Contribution to the Critique of Political Economy*, Marx's work of 1859, by Charles H. Kerr, Chicago, 1904. The quotations in the text above are from pp. 310–12.

theory of history has been contradicted by the twentieth-century clash of different timebound societies.[5]

So we are back to our initial question: what marks off the present from the past, and how do we understand each other; how, for example, do we read the ancient Greeks, and how would they read us? The answer lies, perhaps, in a distinctive interplay of culture and technology. By culture, I mean less than the anthropological view, which includes all "non-material" factors within the framework of a society, but more than the genteel view, which defines culture by some reference to refinement (e.g., the fine arts). By culture, I mean the efforts of symbol makers to define, in a self-conscious way, the *meanings* of existence, and to find some justifications, moral and aesthetic, for those meanings. In this sense, culture guards the continuity of human experience. By technology—in a definition I will expand later—I mean the effort to transform nature for utilitarian purposes. In this sense, technology is always disruptive of traditional social forms and creates a crisis for culture. The ground on which the battle is fought is nature. In this paper, I want to deal largely with the vicissitudes of nature as it is reshaped by technology, and the vicissitudes of technology in its relation to society. To that extent, I have to forgo any extensive discussion of culture, though I shall return to that theme at the end.

II

What Is Nature?

What is nature? Any attempt at specific definition brings one up short against the protean quality of the term. Nature

[5] Socialism has not come as the successor of capitalism. Communist China is technologically more backward than capitalist U.S.A. Is it socially more "progressive"? and on what dimensions: freedom, sexual styles, standard of living, communal care, personal dignity, social cohesion, attachment and loyalty to the country or party or leadership figure? Surely there is no way to "rank" these factors.

is used to denote the physical environment or the laws of matter, the "nature" of man (e.g., his "essence") and the "natural order" of descent (e.g., the family, in botany and society). We talk of "natural selection" as the fortuitous variations in individuals or species which assure survival, and "natural law" as the rules of right reason beyond institutional law.[6] In a satirical passage in *Rasselas*, Samuel Johnson has his young Prince of Abissinia meet a sage who, when asked to disclose the secret of happiness, tells him "to live according to his nature." Rasselas asks the philosopher how one sets about living according to nature and is told a string of generalities that expose the wise man's emptiness.[7]

For my purposes, I restrict the meaning of nature to two usages. The first is what in German—whose fine structure of

[6] As Webster's Second points out: "The conception of *nature* (Gr. *physis*, L. *natura*) has been confused by the mingling of three chief meanings adopted with the word into English, viz.: (1) Creative or vital force. (2) Created being in its essential character; kind; sort. (3) Creation as a whole, esp. the physical universe. The main ambiguity is between *nature* as active or creative and *nature* as passive or created. In the original animistic view, the active vitalistic conception prevailed; but Plato sharply distinguished the passive material from the active formal element, and Aristotle continued the distinction in the conception of a moving cause, or God, as separate from the moved physical universe, or Nature. This antithesis is all but obliterated in pantheistic and naturalistic views. It appears in the pantheism of Spinoza, but the distinction of *natura naturans* and *natura naturata* serves only to discriminate two elements or aspects of the one organic being or substance. The two elements, in the forms of matter and energy, are retained in the modern physical or mechanical view, wherein nature appears as a material universe acting according to rules, but to all intents independent of God or purposive cause." (G. & C. Merriam, Springfield, Mass., 1955), p. 1631.

[7] I am indebted for the illustration to John Wain, from a review of *Sexual Politics* in the London *Spectator*, April 10, 1971. As Mr. Wain writes: "Everyone agrees that happiness comes, and can only come, from living according to nature. And what is that? When woman is assigned a different role from man, is she being thwarted and twisted away from 'nature'? Or is it, on the contrary, the woman who wants to be treated exactly like a man who is turning her back on 'nature' and happiness?"

prefixes allows one to multiply distinctions—is called the *Umwelt*, the organic and inorganic realms of the earth which are changed by man. This is the geography of the world, the environment. The second is what the Greeks called *physis*, or the order of things, which is discerned by man; this natural order is contrasted with *themis*, the moral order, and *nomos*, the legal order. For my purposes, then, nature is a realm outside of man whose designs are reworked by men.[8] In transforming nature, men seek to bring the timeless into time, to bring nature into history. The history of nature, then, is on two levels: the sequential transformations of the *Umwelt* as men seek to bend nature to their purposes, and the successive interpretations of *physis* as men seek to unravel the order of things.[9]

We begin with the *Umwelt*, and with myth. Man remakes nature for the simple and startling reason that man, of all living creatures, "natural man," is not at home in nature. Nature is not fitted to his needs. This is the insight first enunciated by Hesiod in *Works and Days*, and retold by Protagoras in the Platonic dialogues to spell out a moral about human society. The story, of course, is that of Prometheus and Epimetheus. The two brothers, foresight and hindsight,

[8] If nature is outside man, what does one do with the term *human nature?* Despite its ambiguities, it is probably indispensable. Yet, in the effort to keep my distinctions clear, I would use instead the term *human character.*

[9] I realize that I am using the phrase "the history of nature" in a very different way from such physicists (or should one call him a natural philosopher) as C. F. von Weizsäcker, who asserts that nature is historic, if by history one means being *within* time, since all of nature itself is changing—and ten billion years ago there was neither sun nor earth nor any of the stars we know—and, following the theorem of the second law of thermodynamics, events in nature are fundamentally irreversible and incapable of repetition. My history of nature, here, is within the time frame and conceptual map of nature's transformation at the hands of man, and the understandings of nature by man. See C. F. von Weizsäcker, *The History of Nature* (University of Chicago Press, Chicago, 1949).

are charged by the gods with equipping the newly fashioned mortal creatures with "powers suitable to each kind." But, unaccountably—perhaps because of the pride of the younger to excel—Epimetheus asks the older for permission to do the job, and is given the task. He begins with the animals. Some are given strength and others speed, some receive weapons and others camouflage, some are given flight and others means of dwelling underground; those who live by devouring other animals are made less prolific while their victims are endowed with fertility—"the whole distribution on a principle of compensation, being careful by these devices that no species should be destroyed."

But without forethought, Epimetheus squandered all his available powers on the brute beasts, and none was left for the human race. Prometheus, come to inspect the work, "found the other animals well off for everything, but man naked, unshod, unbedded, and unarmed, and already the appointed day had come when man, too, was to emerge from within the earth into the daylight." Prometheus therefore stole from Athena and Hephaestus the gift of skill in the arts, together with fire. "In this way man acquired sufficient resources to keep himself alive. . . ."[10] Nature became refitted for man.

As Prometheus says, in the play of Aeschylus:

I gave to mortals gift.
I hunted out the secret source of fire.
I filled a reed therewith,
fire, the teacher of all arts to men,
the great way through. . . .

I, too, first brought beneath the yoke
great beasts to serve the plow,
to toil in mortals' stead. . . .

10 "Protagoras," in *Plato: The Collected Dialogues*, edited by Edith Hamilton and Huntington Cairns, translated by W. K. C. Guthrie (Bollingen Series LXXI, New York, 1966), lines 320d to 322, pp. 318–19.

Listen, and you shall find more cause for wonder.
Best of all gifts I gave them was the gift of healing.
For if one fell into a malady
there was no drug to cure, no draught, or soothing ointment. . . .

The ways of divination I marked out for them,
and they are many; how to know
the waking vision from the idle dream;
to read the sounds hard to discern;
the signs met on the road; . . .
So did I lead them on to knowledge
of the dark and riddling art.[11]

Natural goods are those we share with the animals, but cultivated or fabricated goods require the reworking of nature: the husbandry of soil and animals, the burning of the forests, the redirection of the rivers, the leveling of mountains. These demand acquired powers. The introduction of *techne* gives man a second nature, or different character, by extending his powers through adaptive skills and redirective thought; it allows him to prefigure or imagine change and then seek to change the reality in accordance with the thought. The fruits of *techne* create a second world, a technical order which is superimposed on the natural order.

In the imagination of the Greeks, these stolen skills were powers of the gods, and with these powers man could begin that ropedance above the abyss which would transform him from "the kinship with the worm," in the phrase of Faust, to the godlike knowledge that partakes of the divine. Prometheus was punished, and, in the romantic imagination of Marx as well as Shelley, Prometheus was the eternal rebel who had dared to act for men. The paradox is that today the romantic imagination, having turned against *techne*, remains puzzled as to what to say about its primal hero. Most likely, the new shamans would say that the punishment was justified. But that is another story, for another day.

[11] Aeschylus, *Prometheus Bound*, translated by Edith Hamilton, in *Greek Plays in Modern Translation*, edited by Dudley Fitts (The Dial Press, New York, 1953), pp. 508–9, 519–20.

I jump now almost two thousand years, from Protagoras to the seventeenth century C.E., to a radically new way of looking at nature and of organizing thought, the rule of abstraction and number.[12]

Mythology, the first mode of depicting the world, is based on personification or metaphor. Nature is a creative or vital force ruling the *Umwelt*. In *Prometheus Bound* the characters are called Ocean, Force, and Violence, or in the later personification of the tides of destiny (we cannot escape metaphor in our speech) we find *Moira*, or Fate, and *Tyche*, or Chance, as the two principles which rule our lives. Through myth, metaphor, and characterization, we can dramatize our plights, and search for meaning in expressive symbolism; that is the virtue of the poetic mode. But with abstraction and number, we can state causal or functional relationships and predict the future states of, or manipulate, the world. Nature as *physis* is an order of things. The heart of the modern discovery is the word *method*. Nature is to be approached through a new method.

In terms of method, the first achievement, that of Galileo, was the simplification of nature. Galileo divided nature into the world of qualities and the world of quantities, the sensory order and the abstract order. All sensory qualities—color, sound, smell, and the like—were classified as secondary and relegated to subjective experience. In the physical world were the primary quantities of size, figure, number, position, motion, and mass, those properties which were capable of

[12] In this section I have drawn primarily from E. J. Dijksterhuis, *The Mechanization of the World Picture* (Oxford University Press, London, 1961); Charles C. Gillespie, *The Edge of Objectivity* (Princeton University Press, Princeton, N.J., 1960); Arthur Koestler, *The Sleepwalkers* (Hutchinson, London, 1959); John Herman Randall, Jr., *The Making of the Modern Mind* (Houghton Mifflin, Boston, 1926), especially for the quotations from Descartes and Spinoza; and Joseph Mazzeo, *Renaissance and Revolution* (Pantheon Books, New York, 1965), especially on Galileo. Unless otherwise noted, the quotations from Descartes and Spinoza are taken from Randall.

extension and mathematical interpretation. The worlds of poetry and physics, the idea of natural philosophy, were thus sundered.

Equally important was the contrast with the classical Aristotelian view which Thomas Aquinas had enlarged upon in medieval thought. Then the object of science was to discover the different purposes of things, their essence, their "whatness," and their qualitative distinction. But little attention was paid to the exactly measured *relations* between events or the *how* of things. In this first break with the past, measurement and relation became the mode. To do so Galileo shifted the focus of attention from specific objects to their abstract properties. One did not measure the fall of an object but mass, velocity, force, as the properties of bodies, and the relations among these properties. The elements of analytical abstraction replaced concrete things as the units of study.

The search for method, which was taken up by Descartes, was not just an effort at exactitude and measurement, but had a double purpose. The first was to raise the general intellectual powers of all men. An artistic ignoramus with a compass, said Descartes, can draw a more perfect circle than the greatest artist working free hand. The correct method would be to the mind what the compass is to the hand. And second, with this "compass of the mind" one could create a general method that would be a flawless instrument for the unlimited progress of the human mind in theoretical and practical knowledge of all kinds.[13]

That "dream of reason" is symbolized by a famous episode

[13] As Descartes wrote:

> . . . instead of that speculative philosophy which is taught in the schools, we may find a practical philosophy . . . by means of which, knowing the force and action of fire, water, air, the stars, heavens and all other bodies that environ us, as distinctly as we know the different crafts of our artisans, we can in the same way employ them in all those uses to which they are adapted, and thus render ourselves the masters and possessors of nature.

The Philosophical Works of Descartes, trans. by Elizabeth S. Haldane and G. R. T. Ross (Cambridge University Press, London, 1931), Vol. I, p. 119.

in Descartes's life. One day, confined to his room by a cold, he resolved to discard all beliefs that could not pass the test of reason. That night he had an intense vision, and with feverish speed he perfected the union of algebra and geometry —the complete correspondence between a realm of abstraction and a realm of real world space—that we now call analytical geometry.

> As I considered the matter carefully, it gradually came to light that all those matters only are referred to mathematics in which order and measurement are investigated, and that it makes no difference whether it be in numbers, figures, stars, sounds or any other object that the question of measurement arises. I saw consequently that there must be some general science to explain that element as a whole which gives rise to problems about order and measurement, restricted as these are to no special subject matter. This, I perceived, was called universal mathematics. . . . To speak freely, I am convinced that it is a more powerful instrument of knowledge than any other that has been bequeathed to us by human agency, as being the source of all others.

Intoxicated by his vision and his success, Descartes declared, "Give me extension and motion, and I will construct the universe." And with Newton's mathematical method of computing the rates of motion, the calculus, the universe was constructed in exact, mathematical deductive terms.

> The universal order [said Newton], symbolized henceforth by the law of gravitation, takes on a clear and positive meaning. This order is accessible to the mind, it is not preëstablished mysteriously, it is the most evident of all facts. From this it follows that the sole reality that can be accessible to our means of knowledge, matter, nature, appears to us as a tissue of properties, precisely ordered, and of which the connection can be expressed in terms of mathematics.

But what of design, purpose, value, *telos?* None. What we have with this "watershed," as Arthur Koestler has called it, is the desacralization, or—depending on one's temperament and values—the demystification, of nature.

The use of mathematics to discover the underlying order of the universe is, of course, not new. The Pythagoreans had sought to discern form, proportion, and pattern, expressed as relations, in the order of number—shapes and intervals which could be expressed in musical terms. The discovery that the pitch of a note depends on the length of the string which produces it and that concordant intervals in the scale are produced by simple numerical ratios, was a reduction of "quality to quantity," or the mathematicization of human experience. But with the Pythagoreans, as it was for Kepler, the mystical and scientific modes of experience were joined, each to illumine the other. With Galileo comes the radical separation. "He was utterly devoid of any mystical, contemplative leanings, in which the bitter passions could from time to time be resolved," Koestler writes; "he was unable to transcend himself and find refuge, as Kepler did in his darkest hours, in the cosmic mystery. He did not stand astride the watershed; Galileo is wholly and frighteningly modern."

With Galileo, physics becomes detached not only from mysticism but from "natural philosophy" as well. Aristotelian physics is valuative, reflecting a world conceived in hierarchic terms. The "highest" forms of motion are circular and rectilinear; and they occur only in the "heavenly" movements of the stars. The "earthly," sublunar world is endowed with motion of an inferior type. But, in the Newtonian world view, the idea of the heavens is detached from any ascending hierarchy of purposes as envisaged by Thomas Aquinas and is but a uniform, mathematical system on the single plane of motion.

And finally, in this mechanistic world view the world becomes sundered from the anthropomorphic image of a wise and loving Father, or the theological image of the being whose powers are so miraculous that he can create a world out of nothing—the doctrine of *creato ex nihilo*. As Spinoza put it in his geometry: "Nothing comes to pass in nature in contravention to her universal laws . . . she keeps a fixed and immutable order." And purpose and final cause? "There is no need to show at length that nature has no particular goal in view, and that final causes are mere human figments. . . . Whatsoever comes to pass, comes to pass by the will and eternal decree of God; what is, whatever comes to pass comes to pass according to rules which involve eternal necessity and truth."

Thus the order of nature, *physis,* is some vast *perpetuum mobile,* whose every point in time, including its future state, can be deduced mathematically from the fundamental principles of its mechanical action. Nature is a machine.

A third vision: man the inventor, the experimenter, the active, purposeful intervenor in the processes of nature to subordinate and bend the *Umwelt* to men's wills. This is a view which centers not on myth or mechanism but on man as an active reshaper of the world. The key word here is *activity.* If there is a radical difference between classical and modern views, it is the abandonment of the contemplative attitude toward nature, aesthetics, and thought, and the adoption of an activist attitude toward experience and the environment. All of modern epistemology is dependent on an activity theory of knowledge.

Our science and technology, Lynn White has written, grew out of the Christian view of nature as the dominion of men over the earth and other creatures. This may be too facile. Christian thought, as Clarence J. Glacken has observed, is also characterized by a *contemptus mundi,* a rejection of the earth as the dwelling place of man and an indifference to nature. And while Christian thought, to follow White, may have felt alien to an idea of a "sacred grove," and could be prodigal in its waste of nature, it did not have the impulse to rework God's designs for man's own purposes.[14]

[14] For Lynn White's view, see "The Historical Roots of Our Ecological Crisis," in *Machina ex Deo* (M.I.T. Press, Cambridge, Mass., 1968), esp. p. 90. "For nearly two millennia Christian missionaries

It is in the period roughly from the end of the fifteenth century to the end of the seventeenth that the sources of the idea of man as a controller of nature and an active agent—in mind and in matter—began to take shape. For our purposes, there are two sources. One is a new and growing emphasis on *practical activity* and the emergence, during the Renaissance, of a group of remarkable men, artist-engineers, who united rational training with manual work to lay down the foundations of experimental science. These artist-engineers, as Edgar Zilsel remarks, not only painted pictures and built cathedrals but also constructed lifting engines, earthworks, canals, sluices, guns, and fortresses, discovered new pigments, formulated the geometrical laws of perspective, and invented new measuring tools for engineering and gunnery. Among these were Filippo Brunelleschi, the principal architect of the cathedral of Florence; the bronze founder Lorenzo Ghiberti; the architect Leon Battista Alberti; the architect and military engineer Francesco di Giorgio Martini; the incomparable Leonardo da Vinci, who drew maps, built canals, created weapons, and designed craft to submerge under the sea and others to fly through the air; the goldsmith, sculptor, gun constructor, and adventurer Benvenuto Cellini; and in Germany, Albrecht Dürer as a surveyor and cartographer. Related to them were instrument makers who supplied navigators, geodesists, and astronomers with their aids, and clock makers, cartographers, and military technicians who created new tools. Their knowledge was empirical, but they sought to systematize and generalize their experiences and recorded them in diaries or treatises for their colleagues and apprentices. From these experiences they sought to discover theories for general application. "Practice in painting must always be founded on sound theory," Leonardo wrote; and Alberti's book *On Painting*, based on geometry and optics, attempted to

---

have been chopping down sacred groves, which are idolatrous because they assume spirit in nature." For a different view see Clarence J. Glacken, *Traces on the Rhodian Shore* (University of California Press, Berkeley and Los Angeles, 1967), Chapter 4, esp. p. 162.

create a new "compass of vision" which would guide all use of perspective in art, as almost two hundred years later Descartes would seek a method for the compass of the mind.[15]

The second source of the "activity principle" derives from the Cartesian revolution in concepts. The contemplative tradition of mind, going back to Greek and Christian thought, saw the human being as an observer passively regarding a world unfolding in front of him; knowledge was a copy, so to speak, of a picture of what that outside world was like. So long as science followed Aristotelian realism, and the world was just what its picture seemed to the mind, there were no difficulties. But Galileo offered a different conception. Water, for example, was not just a wet, formless fluid of varying temperature that one could observe, but a number of particles of matter whose motion followed definite laws. And for Descartes, the qualities of wetness and coolness were not just properties of water, but generalized concepts in the mind that perceived the water. Knowledge of nature, or of the world, thus depended not on immediate experience but on the axioms of geometry and the categories of mind.

The world of appearance and the substructures of reality had come apart. Yet the two could be joined, since what one knows, as Kant developed the idea, is a function of the selective categories by which mind relates different worlds of fact and appearance. Mind organizes perception through selective scanning, takes different attributes and properties of objects or events, groups these together for the purposes of analysis and comparison, and organizes them into conceptual systems whose use is tested by empirical application. Mind is thus an active agent in the making of judgments about

[15] See Edgar Zilsel, "Problems of Empiricism," in *The Development of Rationalism and Empiricism*, International Encyclopedia of Unified Science, Volume II, Number 8 (University of Chicago Press, Chicago, Ill., 1941); Dijksterhuis, *The Mechanization of the World Picture, supra*, pp. 241–47; and, as a case study, Joan Gadol, *Leon Battista Alberti* (University of Chicago Press, Chicago, Ill., 1969).

reality. But these judgments, to be valid, have to be confirmed through the prediction of consequences. To know the world, one has to test it. In the older meanings of the words, *theria* meant to see, and practice, to do; but necessarily, activity joins the two. In this way, rationalism and empiricism are interlocked through *praxis*.

Thus, a new world view: the emphasis on practical activity and on the role of mind as an active formulator of plans reworking the categories of nature. As Descartes observed in Part V of the *Discourse on Method*, what he was seeking for was a science and technique which would make men "the masters and possessors of Nature."

The spread of a world view requires a prophet. The prophet provides the passion for a view; he makes claims (usually extravagant) and gains attention; he codifies the doctrine and simplifies the argument—in short, he provides a formula for quick understanding and a moral rationale for necessary justification. Of this new world view, there were two significant prophets: Francis Bacon and Karl Marx.

Francis Bacon, lawyer, politician, essayist—at one time Lord Chancellor of England in 1618—was not a scientist. As Alexandre Koyré has remarked: If all of Bacon's writings were to be removed from history, not a single scientific concept, not a single scientific result, would be lost. Yet science might have been different without him, or without someone of his literary gifts. For what Bacon did was to formulate the modern credo: of science as the endless pursuit of knowledge; of the experiment as the distinctive mode of science; and of utility as the goal and purpose of science.

Of the endless pursuit of knowledge Edgar Zilsel has observed:

> The modern scientist looks upon science as a great building erected stone by stone through the work of his predecessors and his contemporary fellow-scientists, a structure that will be

> continued but never be completed by his successors. . . . [Science] is regarded as the product of a cooperation for non-personal ends, a cooperation in which all scientists of the past, the present and the future have a part. Today this idea or ideal seems almost self-evident. Yet no Brahmanic, Buddhistic, Moslem or Catholic Scholastic, no Confucian scholar or Renaissance Humanist, no philosopher or rhetor of classical antiquity ever achieved it. It is a special characteristic of the scientific spirit and of modern western civilization. It appeared for the first time fully developed in the works of Francis Bacon.[16]

Against scholasticism, Bacon emphasized the importance of the experiment. He was himself an enthusiastic experimenter, so much so that he died from a cold he caught while stuffing a dead chicken with snow during an experiment on the preservation of food.[17] Bacon believed in the experiment as the foundation of inductive thinking. Science today has a less exalted view of induction, and the most sophisticated philosophies of science mock its possibilities at all; yet the idea of the experiment remains as the necessary condition for the dis-

[16] Edgar Zilsel, "The Genesis of the Concept of Scientific Progress," in Philip Wiener and Aaron Noland (eds.), *Roots of Scientific Thought* (Basic Books, New York, 1960), p. 251.

[17] One might take this activity as an illustration of the range of interest of the man of letters, before "the two cultures." As Franklin Ford observed even of culture a century later:

> Through most of the eighteenth century, a truly learned man, wherever his most highly developed competence might lie, claimed and was conceded the right to move with interest and confidence over a range which we should now describe as covering physical sciences, social sciences, and humanities. What was Montesquieu—political scientist, sociologist, historian, aesthetic philosopher, critic of morals? Before selecting a label, we might reflect that his earliest scholarly prize was won with a paper concerning nervous and muscular reactions observed in the thawing of a frozen sheep's tongue.

"Culture and Communication," unpublished paper for the American Academy of Arts and Sciences conference on *Science and Culture*, Boston, May 10–11, 1963.

confirmation of hypotheses. As Charles C. Gillespie noted: ". . . Bacon's emphasis on experiment did shape the style of science. So strongly did it do so that the term 'experimental science' has become practically a synonym for 'modern science,' and nothing so clearly differentiates post-seventeenth century science from that of the Renaissance, or of Greece, as the role of experiment."[18]

But experiment, for Bacon, was linked to a specific purpose: "The true and lawful goal of the sciences is none other than this: that human life be endowed with new discoveries and powers." As with Leonardo, one finds in Bacon an admiration for the inventor and the experimenter, and a contempt for the pretensions of authority (at least that of the past; at the courts of Elizabeth and James he was a toady and was rewarded for his obsequiousness with a peerage: one can say that he rendered unto science that which was science's and to Caesar that which Caesar demanded; when personal survival is at stake, it is difficult to be harsh on the courtier). It was to the mechanical arts, to industry and seafaring, to the practical men, that Bacon went for the source of knowledge. The scholastics and humanists, he declared, have merely repeated the sayings of the past. Only in the mechanical arts has knowledge been furthered since antiquity. In Bacon's philosophy, there is no choice to be made between basic and applied science. On the contrary, applied science is by definition basic; it is the object of the search.[19]

All of this is summed up in his last book, a utopian fable,

[18] *The Edge of Objectivity*, *supra*, p. 79.

[19] As Charles Gillespie has observed: "His was the philosophy that inspired science as an activity, a movement carried on in public and of concern to the public. This aspect of science scarcely existed before the seventeenth century. . . . This is comfortable democratic doctrine, and it is obvious why Baconianism has always held a special appeal as the way of science in societies which develop a vocation for the betterment of man's estate, and which confide not in aristocracies, whether of birth or brains, but in a wisdom to be elicited from common pursuits—in seventeenth-century England, in eighteenth-century France, in nineteenth-century America, amongst Marxists of all countries." Ibid., p. 75.

The New Atlantis. The choice of form and the selection of place are certainly not fortuitous. The form is a voyage of discovery. Fresh to the times are the great voyages which have expanded man's geographic vision, and for Bacon these voyages symbolize the broadening of intellectual horizons as well. The place is Atlantis, the lost continent of the cosmological allegory of the *Timaeus,* the place where Plato introduced the notion of a demiurge as artisan-deity, the one who creates an orderly universe out of recalcitrant materials.

In Bacon's utopia, the philosopher is no longer king; his place has been taken by the research scientist. The most important building in this frangible land of Bensalem is Salomon's House, not a church but a research institute, "the noblest foundation . . . that ever was upon the earth and the lantern of this kingdom." Salomon's House, or the College of the Six Days Works, was a state institution "instituted for the production of great and marvelous works for the benefit of man." Of the thirty-five pages which make up the complete work, ten are devoted to the listing of the marvels of inventions that have been gathered in this treasure house. The Master of the House, speaking to his visitors, declares: We imitate the flight of birds: for we have some degree of flying in the air; we have ships and boats for going under water. There are engines for insulation, refrigeration and conservation. Animals are made bigger or smaller than their natural size. There is a perspective house to demonstrate light, a sound house for music, an engine house to imitate motions, a mathematical house, and "houses of deceits of the senses, where we represent all manner of . . . false apparitions, impostures, illusions and their failures."

For Seneca (who faced similar political problems) and the Stoics, the exploration of nature was a means of escaping the miseries of life. For Bacon, the purpose of knowledge was utility, to increase happiness and to mitigate suffering. In summing up the marvels, the master of Salomon's House bespeaks the boundless ambition of all technological utopias. "The end of our foundation," he says, "is the knowledge of causes and the secret motions of things . . . the enlarging of the bounds of human empire to the effecting of all things possible."[20]

For Marx, like Protagoras, man, too, is the measure of all things. What is striking is his fierce contempt for the romantic cult of nature and his denigration of the sentimental idylls—"drivel," he called them—about the land and the forests. In *The German Ideology* Marx mocks the utopian socialists who see a harmonious unity in nature. Where the "true socialist" sees "gay flowers . . . tall and stately oaks . . . their satisfactions lie in their life, their growth, their blossoming," Marx remarks, "'Man' could [also] observe . . . the bitterest competition among plants and animals; he could see . . . in his 'forest of tall and stately oaks,' how these tall and stately capitalists consume the nutriment of the tiny shrubs. . . ."[21]

This sarcasm is all the more remarkable since Marx came to his own philosophy through the forest of Feuerbach, his mentor, who preached a sensual cult of nature and proclaimed Man rather than God as the center of the world. In an essay written in 1850, a review of a book by Daumer entitled *The Religion of a New Age*, Marx renounced any kind of Feuerbachian cult of "Man" or "Nature." Where the author tries to establish a "natural religion" in modern form, based on reverence for nature, Marx counterposes science:

> [In the work] there is no question, of course, of modern sciences, which, with modern industry, have revolutionized the whole of nature and put an end to man's childish attitude towards nature as well as to other forms of childishness. . .[22]

[20] Francis Bacon, *New Atlantis*, in *Famous Utopias*, edited by Charles M. Andrews (Tudor Publishing Co., New York, n.d.), pp. 235–72, esp. 263, 269–71.

[21] Quoted in Alfred Schmidt, *The Concept of Nature in Marx* (NLB, London, 1971), pp. 129–30.

[22] Ibid., pp. 131–32.

For Marx, nature is blind, nature is necessity. Against blind nature is conscious man, who, in his growing consciousness, is now able to plan and direct his future. Against necessity, the need to toil, is History, a new demiurge to wrest plenty from a recalcitrant nature. In the movement of History comes the point, in the dithyrambic phrase of Engels, where there is the "leap from the realm of necessity into the realm of freedom," the "genuinely human" period to which all recorded history has been an antechamber. As Marx writes in *Capital:*

> Just as the savage must wrestle with nature to satisfy his wants, to maintain and reproduce life, so must civilized man, and he must do so in all social formations and under all possible modes of production. With his development this realm of physical necessity expands as a result of his wants; but at the same time, the forces of production which satisfy these wants also increase. Freedom in this field can only consist in socialized men, the associated producers, rationally regulating their interchange with nature, bringing it under their common control, instead of being ruled by it as the blind forces of nature. . . But it nonetheless still remains a realm of necessity. Beyond it begins that development of human energy which is an end in itself, the true realm of freedom, which, however, can blossom forth only with this realm of necessity as its basis.[23]

The fulcrum of all this is human needs. Cartesian rationalism and Kantian idealism had created a theory of mind as activity, against the old, passive materialism. But, for Marx,

[23] *Capital,* Volume III, pp. 799–800. There is a subtle difference here between Marx and Engels which is more than of interest to the specialist. In *Anti-Dühring,* which is better known for its formulation of the "leap to freedom" than the passage in *Capital,* Engels assumes that when man "for the first time becomes the real conscious master of nature, because and in so far as he has become master of his own social organization," all necessity, i.e., *all* labor, is abolished. But Marx was less utopian and argued that some necessity would always remain. This question is explored by Alfred Schmidt, op. cit., pp. 134–36.

consciousness alone, though it can interpret, that is, fully prefigure, a world—this is what he meant by the "realization" of philosophy—cannot change the world. Man does so because he has needs which can be satisfied only by transforming nature. With Marx, the activity theory of mind is reworked as an activity theory of material needs and powers.

For Marx, the agency of man's power is technology. History is the history of the forces of production, and the significance of the modern era is the extraordinary power which a new class in history, the bourgeoisie, has achieved through *techne*. In the *Communist Manifesto*, the herald of the socialist revolution, Marx writes a startling panegyric to capitalism:

> The bourgeoisie was the first to show us what human activity is capable of achieving. It has executed works more marvellous than the building of Egyptian pyramids, Roman aqueducts, and Gothic cathedrals. . . .
> During its reign of scarce a century, the bourgeoisie has created more powerful, more stupendous forces of production than all preceding generations rolled into one. The subjugation of the forces of nature, the invention of machinery, the application of chemistry to industry and agriculture, steamships, railways, electric telegraphs, the clearing of whole continents for cultivation, the making of navigable waterways, huge populations springing up as if by magic out of the earth—what earlier generations had the remotest inkling that such productive powers slumbered within the womb of social labour.[24]

[24] D. Ryazanoff (ed.), *The Communist Manifesto of Karl Marx and Friedrich Engels* (Russell & Russell, New York, 1963), pp. 29, 31–32. And there was still more to come. Wilhelm Liebknecht, one of the founders of the German Social Democratic Party, tells of his conversation with Marx when he joined the German Socialist Workers Club in London in 1850: "Soon we were on the field of Natural Science, and Marx ridiculed the victorious reaction in Europe that fancied it had smothered the revolution and did not suspect that Natural Science was preparing a new revolution. That King Steam who had revolutionized the world in the last century had ceased to rule, and that into his place a far greater revolutionist would step, the

"The bourgeoisie cannot exist," wrote Marx, "without incessantly revolutionizing the instruments of production. . . . That which characterizes the bourgeois epoch in contradistinction to all others is a continuous transformation of production." These forces of production, however, are held in check by the property owners, those who can only be superfluous in the new society, and this contradiction sets limits on the development of the economy and the productive forces themselves. Remove the bourgeoisie, and all the brakes on production will be released.

What is clear from all this is Marx's faith in technology as the royal road to utopia; technology had solved the single problem which had kept the majority of men in bondage throughout human history and which was responsible for almost all the ills of the world: economic scarcity. Because of scarcity, each man is pitted against the other, and man becomes wolf to man. Scarcity is the condition of the state of nature, which is why, says Marx, Hobbes was right that the state of nature is a war of one against all.

The "end of history" is the substitution of a conscious social order for a natural order. The unfettered reign of technology is the foundation of abundance, the condition for the reduction, if not the end, of necessity.[25]

---

electric spark. And now Marx, all flushed and excited, told me that during the last few days the model of an electric engine drawing a railroad train was on exhibition in Regent street. 'Now the problem is solved—the consequences are indefinable. In the wake of the economic revolution the political must necessarily follow. . . .'" Wilhelm Liebknecht, *Karl Marx: Biographical Memoirs* (Charles H. Kerr, Chicago, 1901), p. 57.

[25] Can there ever be an end to scarcity? What is striking is that Marx, like all nineteenth-century thinkers, conceived of abundance as a plethora of goods. But, in contemporary society, there are many new—and crucial—kinds of scarcity which were never envisaged, notably the scarcity of time. And if we think of abundance, as an economist does, as items being a "free good" (i.e., as having a zero cost, as clean air once had), then one finds in contemporary society

III

What Is Technology?

Technology, like art, is a soaring exercise of the human imagination. Art is the aesthetic ordering of experience to express meanings in symbolic terms, and the reordering of nature—the qualities of space and time—in new perceptual and material form. Art is an end in itself; its values are intrinsic. Technology is the instrumental ordering of human experience within a logic of efficient means, and the direction of nature to use its powers for material gain. But art and technology are not separate realms walled off from the other. Art employs *techne*, but for its own ends. *Techne*, too, is a form of art that bridges culture and social structure, and in the process reshapes both.[26]

This is to see technology in its essence. But one may understand it better, perhaps, by looking at the dimensions of its existence. For my purposes, I will specify five dimensions in order to see how technology transforms both culture and social structure.

1. *Function.* Technology begins with an aesthetic idea: that the shape and structure of an object—a building, a vehicle, a machine—are dictated by its function. Nature is a guide,

---

that a host of new activities, such as information, or the co-ordination of activities in participatory situations, create rising costs and therefore become increasingly scarce resoures. For an elaboration of this argument, see the section "The End of Scarcity" in my book *The Coming of Post-Industrial Society* (Basic Books, New York, 1973).

[26] *Techne,* it should be noted, though we use it to denote technology, in Greek means art, both in the sense of a craft used by a craftsman (artisan), and as Aristotle defined it, as consisting "in the conception of the result to be produced before its realization in the material." ("De Partibus Animalium," 641, in *The Basic Works of Aristotle,* ed. by Richard McKeon (Random House, New York, 1941), pp. 647–48). I am indebted to Emanuel Mesthene for the suggestion.

only to the extent that it is efficient. Design and form are no longer ends in themselves. Tradition is no justification for the repetition of designs. Form is not the unfolding of an immanent aesthetic logic—such as the musical forms of the eighteenth to early-twentieth centuries. There is no dialogue with the past. It is no accident that the adherents of a machine aesthetic in the early twentieth century, in Italy and Russia, flaunted the name Futurism.

2. *Energy*, Technology is the replacement of natural sources of power by created power beyond all past artistic imagination. Leonardo made designs for submarines and air-conditioning machines, but he could not imagine any sources of power other than what his eyes could behold: human muscles or animal strength, the power generated by wind and falling water. Visionaries of the seventeenth century talked grandiosely of mechanized agriculture, but their giant combines were to be driven by windmills and thus could not work. Energy drives objects—ships, cars, planes, lathes, machines—to speeds thousands of times faster than the winds, which were the limits of the "natural" imagination; creates light, heat, and cold, extending the places where people can live and the time of the diurnal cycle; lifts weights to great heights, permitting the erection of scrapers of the skies and multiplying the densities of an area. The skyscraper lighted at night is as much the technological symbol of the modern city as the cathedral was the emblem of medieval religious life.

3. *Fabrication*. In its oldest terms, technology is the craft or scientific knowledge which specifies ways of doing things in a reproducible manner.[27] The replication of items from templates or dies is an ancient art; its most frequent example is coinage. But the essence of technology is that, owing to the standardization of skills and the standardization of objects, its reproduction is much cheaper than its invention or development. Modern technological fabrication introduces two different factors: the replacement of manual labor and artisan skills by programmed machines; and the incredible

[27] I owe this definition to Harvey Brooks.

rapidity of reproduction—the printing of a million newspaper copies per night—which is the difference in scale.

4. *Communication and Control.* Just as no one before the eighteenth century could imagine the new kinds of energy to come, so, well into the nineteenth century, could no one imagine—even with the coding of messages into dots and dashes, as in telegraphy—the locking of binary digits with electricity or the amplification of ethereal waves, which has produced modern communication and control systems. With telephone, radio, television, and satellite communication, one person can talk to another in any part of the globe, or one person can be seen by hundreds of millions of persons at the same moment. With programmed instructions—through the maze of circuits at nanoseconds speed—we have control mechanisms that switch trains, guide planes, run automated machinery, compute figures, process data, simulate the movement of the stars, and correct for error both human and machine. And an odd phrase sums it all up: these are all done in "real time."

5. *Algorithms.* Technology is clearly more than the physical manipulation of nature. There is an "intellectual technology" as well. An algorithm is a "decision rule," a judgment of one or another alternative course to be taken, under varying conditions, to solve a problem. In this sense we have technology whenever we can substitute algorithms for human judgment.

We have here a continuum with classical technology, but it has been transposed to a new qualitative level. Physical technology—the machine—replaced human power at the *manual* level of raw muscle power or finger dexterity or repetition of tasks; the new intellectual technology—as embodied in a computer program or a numerically controlled machine tool—substitutes algorithms for human *judgments*, where these can be formalized. To this extent, the new intellectual technology marks the last half of the twentieth century, as the machine was the symbol of the first half.

Beyond this is a larger dream, the formalization of a theory of choice through stochastic, probabilistic, and deterministic methods—the applied mathematics of Markov chains, Monte Carlo randomization, or the minimax propositions of game theory. If the computer is the "tool," then "decision theory" is its master. Just as Pascal sought to throw dice with God, or the physiocrats to draw a perfect grid to array all economic exchanges among men, so the decision theorists, and the new intellectual technology, seek their own *tableau entier*—the compass of rationality itself.

We say knowingly that technology has transformed the world, and we feel, apprehensively, that we are on a steeply rising "exponential curve" of change, so that the historical transition to new levels of technological power all over the world creates a crisis of transformation. But it is not always clear what has been transformed. To say vaguely "our lives," or flashily that we are experiencing "future shock," is of little help in understanding the character of that transformation. Even the fashionable phrase "the acceleration in the pace of change" is of little help, for it rarely specifies change in *what*. If one thinks of inventions that change the character of daily life, it is not at all clear that our generation is experiencing "more" change (and how does one measure it?) than previous ones. As Mervyn Jones points out, "A man born in 1800 and dying in 1860 would have seen the coming of the railway, the steamship, the telegraph, gas lighting, factory-made clothing and furniture. A man born in 1860 and dying in 1920 would have seen the telephone, electric light, the car and the lorry, the aeroplane, radio and the cinema."[28] Are television and the computer any more "shocking" in their impact on our lives than the inventions which changed the lives of our immediate forebears?

Yet technology has transformed our lives in more radical, yet more subtle, forms than the marvelous "gadgets" that we can see. From a sociological point of view, the effects are

[28] Mervyn Jones, "Tomorrow Is Yesterday," *New Statesman*, Oct. 20, 1972.

twofold: a change in the "axis" of economics, and an increase in the interaction—and consequently the "moral density," to use Durkheim's phrase—among men. These two constitute a definition of "modernity."

For most of the thousands of years of human existence, the level of production dictated the levels of consumption; in the shorthand of economics, supply shaped demand. The returns from nature were small, and, from any single unit of production—in almost all cases, land—they were also diminishing, so that after a period of time men would either have to wait and let the forests and land lie fallow, to be replenished slowly by nature, or move on in search of new lands. Given the scarce returns men fought all the more bitterly for their share, and the major means of amassing wealth were war and plunder. With technology, this situation was changed. One finds an increasing proportion of returns for a given effort—we call this productivity—and, more subtly, a shift in the axis of economics from supply to demand. What men now want begins to affect the levels of production and to dictate the different kinds of items to be produced. It is this change from supply to demand which, in the intellectual sphere, creates modern economics. In the social world, it creates an entirely new attitude towards the world, wealth, and happiness; the three become defined by the sharp rise in the standard of living of the masses of men in the world.[29]

The second change is the breakdown of segmentation and the enlargement of the boundaries of society—the creation, in sociological terms, of the mass society. The mass society is not just a large society. Czarist Russia and Imperial China were large-land-mass societies, but they were segmented; each village was much like any other, repeating

[29] The extraordinary increase in world-wide demand in the past decade has led to new shortages in food, materials, and energy. That these new shortages lead to higher prices goes without saying. Whether these shortages will persist, i.e., that no new sources of materials, energy, or intensive increase in food yields will be found, is moot. But the major point remains: it is the emphasis on demand, an economic fact only a century old, that has led to this state of affairs.

within its bounds the same kind of social structure. Segmentation disappears not with the growth of population but with the increase in the degree of social intercourse or contact—or, more technically, the rate of interaction, per unit of time, between men. Physical density is the number of persons per unit of space. Sociological density is the number of connections among people at a given time. When the Constitution was first adopted, the population of the United States was four million, and New York City, then the capital, held a total of 30,000 persons. Today there are more than 200 million persons in the country, and the metropolitan areas within which people live and work commonly hold from 5 to 20 million persons. Yet the change in the country is not just the increase in numbers, but the quantum jump in interactions as well. As against the life of individuals at the turn of the nineteenth century, if we think of how many different individuals we meet (simply to speak to), how many individuals we know (those we encounter socially), and how many we know of (in order to recognize their names and be responsive to some comment about them), we get some sense of the change in the scale of our lives.

This axial change in the character of social relations brings with it two contrary changes in the areas of work and ideas: differentiation in the social structure and syncretism in the culture.

In the social structure—the realm of the economy, occupations, and stratification system of the society—the increase in interaction (between persons, firms, cities, regions) leads inevitably to competition. In earlier times, with few resources and growth limited, such a zero-sum situation—in which one could gain only at the expense of the other—led to plunder and exploitation. But, in an economic world where everybody can gain (albeit in differential amounts), competition leads to specialization, differentiation, and interdependence. "We can say," Durkheim remarked, "that the progress of the division of labor is in direct ratio to the moral or dynamic density of the society." The condition of efficiency, which is the basis of economic progress, is the growth of specialization and the narrower focusing of tasks and skills. In the area of work, intellectual and otherwise, man becomes a smaller and smaller part of a larger and larger whole. He becomes defined by his role.

In culture, the situation is remarkably reversed. In the traditional society the ideas one has, the beliefs one accepts, the arts one beholds are all within a bounded space. Modernity bursts the walls. Everything is now available. Hindu mantras and Tantric mandalas, Japanese prints and African sculptures, Eskimo music and Indian ragas all jostle with one another in "real time" within the confines of Western homes. Not since the age of Constantine has the world seen so many strange gods mingling in the meditative consciousness of the middle-class mind.

In principle, much of this is not new. What is distinctive is the change of scale. If any single principle dominates our life, it is that. All that we once knew played out on the scale of the Greek polis is now played out in the dimensions of the entire world. Scale creates two effects: One, it extends the range of control from a center of power. (What is Stalin, an unknown wit remarked, if not Genghis Khan with a telephone?) And two, when linear extension reaches certain thresholds, unsettling changes ensue. A university of fifty thousand students may bear the same name that university had thirty years before with five thousand students, but it is no longer the same university. A change in quantities is a change in quality; a change in scale is a change in institutional form.

This principle was laid down more than three hundred and fifty years ago by Galileo, in the square-cube law. As something doubles in size, its volume will triple; but then its shape will also be different. As the biologist D'Arcy Wentworth Thompson pointed out:

> [Galileo] said that if we tried building ships, palaces or temples of enormous size, beams and bolts would cease to

> hold together; nor can Nature grow a tree nor construct an animal beyond a certain size while retaining the proportions and employing the materials which suffice in the case of a smaller structure. The thing will fall to pieces of its own weight unless we either change its relative proportions . . . or else we must find a new material, harder and stronger than was used before.[30]

The major question which confronts the twenty-first century is the question of the limits of scale. The technological revolution, as I have indicated, consisted in the availability of huge amounts of energy at cheaper cost, more control of the circumstances of production, and faster communication. Each development increased the effectiveness of the other two. All three factors increased the speed of performing large-scale operations. Yet, as John von Neumann pointed out seventeen years ago:

> . . . throughout the development, increased speed did not so much shorten time requirements of processes as extend the areas of the earth affected by them. The reason is clear. Since most *time* scales are fixed by human reaction times, habits, and other physiological and psychological processes, the effect of the increased speed of technological processes was to enlarge the *size* of units—political, organizational, economic, and cultural—affected by technological operations. That is, instead of performing the same operations as before in less

[30] *On Growth and Form* (Cambridge University Press, London, 1963) (1st edition, 1917), p. 27. Taking the story of Jack the Giant Killer, Thompson pointed out that Jack had nothing to fear from the Giant. If the Giant were ten times as tall as a man but built like one, he was a physical impossibility. According to the square-cube law, the Giant's mass would be 10^3, or a thousand times Jack's, because he was ten times as big in every dimension. However, the cross section of his leg bones, if he had the shape of a man, would have increased only in two dimensions, 10^2, or a hundred times as big as Jack. A human bone will not support ten times its normal load, and if the Giant tried to walk he would break his legs. Jack was perfectly safe.

> time, now larger-scale operations were performed in the same time. This important evolution has a natural limit, that of the earth's actual size. The limit is now being reached, or at least closely approached.[31]

We first reached that limit in the geopolitical military sphere. In previous epochs, geography could provide an escape. Both Napoleon and Hitler became bogged down in the large land mass of Russia, though by 1940 even the larger countries of continental Western Europe were inadequate as military units. Since 1945, and particularly with the development of intercontinental ballistic missiles with multiple warheads, space and distance offer no effective cover or retreat in any part of the earth. As Von Neumann presciently observed in 1955, "The effectiveness of offensive weapons is such as to stultify all plausible defensive time scales." To recall an old phrase of Winston Churchill's, the equilibrium of power has become the balance of terror.

We are told that we will soon be reaching the limits of resources on a world scale. As a theorem, this is a tautology; as a practical fact, the time scale is elusive. The studies that have sounded the warning are faulty in their methodology.[32] They take little account of economics and the principles of relative prices and substitutability. The resources that may be available from untapped areas (e.g. the ocean bottoms, the large land areas of the Amazon, central Asia, Siberia, western China, the Antarctic) are uncharted. The degree of technological innovation that allows us to substitute lighter and cheaper materials for heavy and cumbersome ones (e.g. the role of semiconductors such as transistors in radios, television sets, and computers)—the entire range of miniaturization—is an unknown factor. And yet the warning is useful. More than

[31] "Can We Survive Technology?" *Fortune*, June 1955, reprinted in *The Fabulous Future* (Dutton, New York, 1956), p. 34.

[32] See, for example, D. H. Meadows et al., *The Limits to Growth* (New American Library, New York, 1972), and, for a telling critique, Carl Kaysen, "The Computer That Printed Out W*O*L*F," *Foreign Affairs*, July 1972.

two thirds of the world is in a pre-industrial phase wherein 60 per cent or more of the labor force is engaged in agriculture, timber, fishing, or mining, games against nature whose returns inevitably diminish. These countries obviously want to industrialize in order to raise their standards of living. The question we must confront is whether our resources are sufficient for such a task, whether new technologies can provide a more economical way, or whether the new industrialization and standard of living must come from some redistribution of the wealth of the advanced industrial countries of the world.

It is possible that we are reaching that limit of scale in technological terms. In the last century, we have increased our speeds of communication by a factor of 10^7, our speeds of travel by 10^2, our speeds of computer operation by 10^6, and our energy resources by 10^3.[33] But all exponential growth reaches an asymptote, the ceiling limit, where it levels off. In terrestrial speed, there is a natural limit of sixteen thousand miles an hour, since any higher speed throws a vehicle out of the earth's orbit. With aircraft, we are questioning whether we should go above supersonic speed because of the danger it might present to the earth's atmosphere or to human noise tolerance on the ground. In communication around the world, we have already approached, in telephonic, radio, and television communication, "real time," and the technological problems are primarily those of expanding the number of bands of communication to permit more and more people to enjoy that use.

In a fundamental sense, the space-time framework of the world *oikoumene* is now almost set. Transportation and communication bind the world as closely together today as the Greek polis of twenty-five hundred years ago. The major sociological problem created by that technology is what happens when all segmentation breaks down and a quantum jump in human interaction takes place. How will we manage when each and every part of the globe becomes accessible to

[33] I take these figures from an essay by John Platt, "What We Must Do," *Science*, Vol. 166, Nov. 28, 1969, pp. 1115–21.

every person? It was once suggested that 7×720 (5,040) citizens was the optimum size for the city-state. (If half a day a year is needed to maintain contact with a relatively good friend, there is a ceiling of 720 persons with whom we could have personal interaction.) Athens, the largest of the ancient Greek city-states, had, at the highest estimate, forty thousand male citizens, and a quorum in the assembly was fixed at six thousand. The number of adult citizens of New Zealand is around 30 times that in Athens, of the Netherlands 100 times, of France 500, of the United States about 2,500, and of India, the largest representative democracy in the world, about 5,000 or 6,000 times.[34] In the face of these numbers, what does participation mean? What is the character of human contact? What are the limits of human comprehension?

IV

What Is Society?

The rhetoric of apocalypse haunts our times. Given the recurrence of the Day of Wrath in the Western imagination—when the seven seals are opened and the seven vials pour forth—it may be that great acts of guilt provoke fears of retribution which are projected heavenward as mighty punishments of men. A little more than a decade ago we had the apocalyptic specter (whose reality content was indeed frightening) of a nuclear holocaust, and there was a flood of predictions that a nuclear war was a statistical certainty before the end of the decade. That apocalypse has receded, and other guilts produce other fears. Today it is the ecological crisis, and we find, like the drumroll of Revelation 14 to 16 recording the plagues: *The Doomsday Book*, *Terracide*, *Our*

[34] I take these examples from Martin Shubik. "Information, Rationality and Free Choice," in Daniel Bell (ed.), *Toward the Year 2000* (Houghton Mifflin, Boston, 1968); and Robert A. Dahl, "The City in the Future of Democracy," *American Political Science Review*, December 1967.

Plundered Planet, The Chasm Ahead, The Hungry Planet, and so on.[35]

In the demonology of the time, "the great whore" is technology. It has profaned Mother Nature, it has stripped away the mysteries, it has substituted for the natural environment an artificial environment in which man cannot feel at home.[36] The modern heresy, in the thinking of Jacques Ellul, the French social philosopher whose writing has been the strongest influence in shaping this school of thought, has been to enshrine *la technique* as the ruling principle of society.

Ellul defines technique as:

> the translation into action of man's concern to master things by means of reason, to account for what is subconscious, make quantitative what is qualitative, make clear and precise

[35] The temper is not restricted to ecologists. Alfred Kazin cites the titles of some recent cultural-social analyses of "our situation": *Reflections on a Sinking Ship, Waiting for the End, The Fire Next Time, The Economy of Death, The Sense of an Ending, On the Edge of History, Thinking About the Unthinkable.*

[36] Theodore Roszak, for example, writes: ". . . we must not ignore the fact that there *is* a natural environment—the world of wind and wave, beast and flower, sun and stars—and that preindustrial people lived for millennia in close company with that world, striving to harmonize the things and thoughts of their own making with its non-human forces. Circadian and seasonal rhythms were the first clock people knew, and it was by co-ordinating these fluid organic cycles with their own physiological tempos that they timed their activities. What they ate, they had killed or cultivated with their own hands, staining them with the blood or dirt of their effort. They learned from the flora and fauna of their surroundings, conversed with them, worshiped them, and sacrificed to them. They were convinced that their fate was bound up intimately with these non-human friends and foes, and in their culture they made place for them, honoring their ways."

What is striking in this evocation of a pagan idyl is the complete neglect of the diseases which wasted most "natural" men, the high infant mortality, the painful, frequent childbirths which debilitated the women, and the recurrent shortages of food and the inadequacies of shelter which made life nasty, brutish, and short.

> the outlines of nature, take hold of chaos and put order into it.

Technique, by its power, takes over the government:

> Theoretically our politicians are at the center of the machinery, but actually they are being progressively eliminated by it. Our statesmen are important satellites of the machine, which, with all its parts and techniques, apparently functions as well without them.

Technique is a new morality which "has placed itself beyond good and evil and has such power and autonomy [that] it in turn has become the judge of what is moral, the creator of a new morality." We have here a new demiurge, an "unnatural" and "blind" logos that in the end enslaves man himself:

> When technique enters into the realm of social life, it collides ceaselessly with the human being. . . . Technique requires predictability and, no less, exactness of prediction. It is necessary, then, that technique prevail over the human being. For technique, this is a matter of life and death. Technique must reduce man to a technical animal, the king of the slaves of technique.[37]

[37] Jacques Ellul, *The Technological Society* (Knopf, New York, 1964), Chapter II, *passim*. What is striking in this unsparing attack on technique is Ellul's omission of any discussion of nature, or how man must live without technique. (The word *nature* does not appear in the index, and there are only a few passing references to the natural world, e.g. p. 79.) As Ellul's translator, John Wilkinson, writes in the Introduction: "In view of the fact that Ellul continually apostrophizes technique as 'unnatural' (except when he calls it the 'new nature'), it might be thought surprising that he has no fixed conception of nature or the natural. The best answer seems to be that he considers 'natural' (in the good sense) *any* environment able to satisfy man's material needs, *if* it leaves him free to use it as means to achieve his individual internally generated ends." Ibid., p. xix.

Ellul has painted a reified world in which *la technique* is endowed with anthropomorphic and demonological attributes. (Milton's Satan, someone remarked, is Prometheus with Christian theology.) Many of the criticisms of technology today remind one of Goethe, who rejected Newton's optics on the ground that the microscope and telescope distorted the human scale and confused the mind. The point is well taken, if there is confusion of realms. What the eye can see unaided, and must respond to, is different from the microcosm below and the macrocosm beyond. Necessary distinctions have to be maintained. The difficulty today is that it is the critics of technology who absolutize the dilemmas and have no answers, short of the apocalyptic solutions that sound like the familiar comedy routine "Stop the world, I want to get off."

Against such cosmic anguish one feels almost apologetic for mundane answers. But after the existentialist spasm, there remain the dull and unyielding problems of ordinary, daily life. The point is that technology, or technique, does not have a life of its own. There is no immanent logic of technology, no "imperative" that must be obeyed. Ellul has written: "Technique is a means with a set of rules for the game. . . . There is but one method for its use, one possibility."[38]

But this is patently not so if one distinguishes between technology and the social "support system" in which it is embedded.[39] The automobile and the highway network form a technological system; the way this system is used is a question of social organization. And the relation between the two can vary considerably. We can have a social system that emphasizes the private use of the automobile; money is then spent to provide parking and other facilities necessary to that purpose. On the other hand, arguing that an automobile is a capital expenditure whose "down time" is quite large, and that

[38] Ibid., p. 97,

[39] The distinction is made in the report of the National Academy of Sciences, *Technology: Processes of Assessment and Choice*, published by the Committee on Science and Astronautics, U. S. House of Representatives, July 1969. See p. 16.

twenty feet of street space for a single person in one vehicle is a large social waste, we could penalize private auto use and have only a rental and taxi system that would substantially reduce the necessary number of cars. The same technology is compatible with a variety of social organizations, and we choose the one we want to use.

One should also distinguish between technology and the accounting system that allocates costs. Until recently, the social costs generated by different technologies have not been borne by the individuals or firms responsible for them, because the criterion of social accountability was not used. Today that is changing. The technology of the internal-combustion engine is being modified because the government now insists that the pollution it generates be reduced. And the technology is being changed. The energy crisis we face is less a physical shortage than the result of new demands—by consumers, and by socially minded individuals for a different kind of technological use of fuels. If we could burn the high-sulphur fuels used until a few years ago, there would be less of an energy crisis; but there would be more pollution. Here, too, the problem is one of costs and choice.

The source of our predicament is not the "imperatives" of technology but a lack of decision mechanisms for choosing the kinds of technology and social support patterns we want. The venerated teacher of philosophy at City College Morris Raphael Cohen used to pose a question to his students in moral philosophy: If a Moloch God were to offer the human race an invention that would enormously increase each individual's freedom and mobility, but demanded the human sacrifice of thirty-thousand lives (the going price at the time), would you take it? That invention, of course, was the automobile. But we had no mechanisms for assessing its effects and planning for the control of its use. Two hundred years ago, no one "voted" for our present industrial system, as men voted for a polity or a constitution. To this extent, the phrase "the industrial revolution" is deceptive, for there was no single moment when people could decide, as they did politically in 1789 or 1793 or 1917, for or against the new system. And yet today, with our increased awareness of alternates and consequences, we are beginning to make those choices. We can do this by technology assessment, and by social policy which either penalizes or encourages a technological development (e.g., the kind of energy we use) through the mechanism of taxes and subsidy.

A good deal of our intellectual difficulty stems from the way we conceive of society. Émile Durkheim, one of the founding fathers of modern sociology, contributed to this difficulty by saying that society exists *sui generis*, meaning that it could not be reduced to psychological factors. In a crucial sense he was right, but in his formulation he pictured society as an entity, a collective conscience outside the individual, acting as an external constraint on his behavior. And this lent itself to the romantic dualism of the individual versus society.

Society is *sui generis*, a level of complex organizations created by the degree of interdependence and the multiplicity of ties among men. A traffic jam, as Thomas Schelling has pointed out, is best analyzed not in terms of the individual pathologies of the drivers, but by considering the layout of roads, the pattern of flow into and out of the city, the congestion at particular times because of work scheduling, and so on. Society is not some external artifact, but *a set of social arrangements, created by men*, to regulate normatively the exchange of wants and satisfactions.

The order of society differs from the order of nature. Nature is "out there," without *telos*, and men must discern its binding and constraining laws to refit the world. Society is a moral order, defined by consciousness and purpose, and justified by its ability to satisfy men's needs, material and transcendental. Society is a design that, as men become more and more conscious of its consequences and effects, is subject to reordering and rearrangement in the effort to solve its quandaries. It is a social contract, made not in the past but in the present, in which the constructed rules are obeyed if they seem fair and just.

The problems of modern society arise from its increasing complexity and interdependence—the multiplication of interaction and the spread of syncretism—as old segmentations break down and new arrangements are needed. The resolution of the problem is twofold: to create political and administrative structures that are responsive to the new scales, and to develop a more comprehensive or coherent creed that diverse men can share. The prescription is easy. It is the exegesis, as the listener to Rabbi Hillel finally understood, that is difficult.[40]

V

The Resolution of Realms

I return to my original question: Is the evident expansion of man's powers a measure of progress; and how do we talk to the Greeks and they to us? I began this discussion with the myth as told by Protagoras, but I did not finish it then; we now return to it.

Following the theft of fire, "man had a share in the portion of the Gods." But he soon found that *techne* does not create civilized life. When men gathered in communities, they injured one another for want of political skill. As Protagoras recounts it:

> Zeus therefore, fearing the total destruction of our race, sent Hermes to impart to men the qualities of respect for others and a sense of justice, so as to bring order into our cities and create a bond of friendship and union.
>
> Hermes asked Zeus in what manner he was to bestow

[40] The traditional story is told that an impatient man once asked Rabbi Hillel to tell him all there was in Judaism while standing on one foot. The Rabbi pondered, and replied: "Do *not* do unto others as you would *not* have them do unto you. All the rest is exegesis."

> these gifts on men. "Shall I distribute them as the arts were distributed—that is, on the principle that one trained doctor suffices for many laymen, and so with the other experts? Shall I distribute justice and respect for their fellows in this way, or to all alike?"
>
> "To all," said Zeus. "Let all have their share. There could never be cities if only a few shared in these virtues, as in the arts. Moreover, you must lay it down as my law that if anyone is incapable of acquiring his share of these two virtues he shall be put to death as a plague to the city."[41]

What we have here, in Homer's earlier terms, is the contrast between *techne* and *themis*. *Techne* enables us to conquer nature; it is essential to the *economic* life. But *themis*, the marriage of conscience and honorable conduct, is the principle of civilized life. "*Themis*," as James Redfield puts it, "is the characteristic human good, and man is distinguished from the feral savage by his ability to live in a society." In Homer, *themis is* primary, and *techne* secondary.[42]

Hegel interpreted Plato's myths as representing a necessary stage in the education of the human race—the childhood phase—which conceptual knowledge could discard as soon as philosophy had grown up. But it should be clear by now that the image of childhood, as used by Hegel and Marx, is meaningless. We are not much ahead of the Greeks in the formulation of our problems or in our wisdom for solving them. In what sense, then, are we alike, and in what sense different?

Society, I would say, should be regarded as having three analytically distinct dimensions—the culture, the polity, and the social structure—each characterized by a different axial principle and each possessing a different historical rhythm.

Culture embraces the areas of expressive symbolism (painting, poetry, fiction), which seeks to explore these meanings

[41] Protagoras, op. cit., sections 322 c–d, pp. 319–20.

[42] James Redfield, "The Sense of Crisis," in John R. Platt (ed.), *New Views of the Nature of Man* (University of Chicago Press, Chicago, Ill., 1965), p. 122.

in imaginative form; the codes of guidance for behavior, which spell out the limits, prescriptive and prohibitive, of moral conduct; and the character structure of individuals as they integrate these dimensions in their daily lives. But the themes of culture are the existential questions that face all human beings at all times in the consciousness of history—how one meets death, the nature of loyalty and obligation, the character of tragedy, the definition of heroism, the redemptiveness of love—and there is a principle of limited possibilities in the modes of response. The principle of culture, thus, is a *ricorso*, returning, not in its forms but in its concerns, to the same *essential* modalities that represent the finitude of human existence.

The polity, which is the regulation of conflict under the constitutive principle of justice, involves the different forms of authority by which men seek to rule themselves: oligarchy and democracy, elite and mass, centralization and decentralization, rule and consent. The polity is *mimesis*, in which the forms are known and men choose those appropriate to their times.

The social structure—the realm of the economy, technology, and occupational system—is *epigenetic*. It is linear, cumulative, and quantitative, for there are specific rules for the process of growth and differentiation.

To the extent that man becomes more and more independent of nature, he can choose and construct the kind of society he wants. Yet he is constrained by the axial facts that each societal realm has a different rhythm of change and that not all forms are compatible with each other.

If one asks, then, in what ways we have advanced beyond the Greeks, we know that our time-space perceptions of the earth have changed, for we have both speed and the view from the air, which the Greeks never knew. And in the power to transform nature and extend the range of man's political life, we live on a scale they would not have been able to understand. Our social structures, then, are vastly different, transformed as they are by technology. Our polities resemble each other in their predicaments (one has only to read Thucydides to be struck by the resemblances), but the problems today are greatly distended by the influence of numbers of persons, and the simultaneity of issues. Yet when we read the major chorus of Sophocles' *Antigone*—"Wonders are many," ending with the antistrophe "the craft of his engines has passed his dream / In haste to the good or evil goal"—we know that, with all the celebration of man's powers to navigate the seas and to domesticate the earth, man without justice and righteousness ("no city hath he who, for his rashness, dwells with sin") is his own enemy, and that we are all, over the span of the millennia, human, all too human.

Coda

In the *ricorsi* of human existence, there have been recurrent cycles of optimism and despair. In the Greek world one finds Hesiod regarding society as corrupt, nature as recalcitrant, and history as regressive, since the golden age lay in the past. For Pericles, some centuries later, society is open, nature is malleable, and history is progressive. But by the end of that century, by the time of Euripides, society is seen as a weak illusion, nature a harsh reality, and history as meaningless.[43] The modern world has had its own cycles. At the dawn of modernity, Rousseau saw society as repressive, nature as good, and history as an illusion. Less than a hundred years later, Comte saw society as open, nature as malleable, and history as progressive. Today the cultural pessimists see society as a monster, nature as recalcitrant, and history as apocalyptic. Is this to be an endless recurrence?

The history of consciousness suggests a resolution. The uniqueness of man lies in his capacity, for self-consciousness and self-transcendence, to stand continually "outside" himself and to judge himself. This is the foundation of human freedom. It is this radical freedom which defines the glory and the plight of man. The modern view of man takes over only

[43] Ibid., pp. 128, 135, 142.

the aspect of freedom, not his finitude; it sees man as a creature of infinite power able to bend the world to his own will: Nothing is unknowable, Comte declared; Prometheus is my hero, Marx declared; Man can make himself, modern humanist psychology declares. It is man's incorrigible tendency towards self-aggrandizement, self-infinitization, and self-idolation which, in the political religions, becomes a moralizing absolute and, owing to the intrinsic egoism of human nature, masks a will to power.

Modern culture, particularly in its utopian versions, denies the biblical idea of sin. Sin derives from the fact that man as a limited and finite creature denies his finiteness and seeks to reach beyond it—beyond culture, beyond nature, beyond history. Evil, as Reinhold Niebuhr has put it, does not exist in nature, but in human history: ". . . human freedom breaks the limit of nature, upsetting its limited harmonies and giving a demonic dimension to its conflicts. There is therefore progress in human history; but it is a progress of all human potencies, both for good and evil."[44]

Thus there is a dual aspect to man as he stands recurrently at the juncture of nature and history. As a creature of nature, he is subject to its brutal contingencies; as a self-conscious spirit, he can stand outside both nature and history and strive to establish his own freedom, to control the direction of his fate. But human freedom is a paradox. Man is limited, subject to causal necessity, and bound to finite conditions; yet, because of his imagination, he is free to choose his own future and be responsible for his own actions. He is able to step over his own finiteness, yet that very step itself risks sin because of the temptations of idolatry—particularly of the will to power. That is the contradiction between finitude and freedom. That is the quandary of human existence.

[44] Reinhold Niebuhr, *The Nature and Destiny of Man* (Charles Scribner's Sons, New York, 1945).

DANIEL BELL

Daniel Bell, professor of sociology at Harvard University, is best known for his wide-ranging and original essays of social criticism. He is the author, among other works, of *The End of Ideology* (1960), *The Reforming of General Education* (1965), and *The Coming of Post-Industrial Society* (1973).

He is vice-president (for the social sciences) of the American Academy of Arts and Sciences, and chairs its Commission on the Year 2000. Before coming to Harvard in 1969, Bell had served on the faculties of the University of Chicago and Columbia University, and was a Fellow at the Center for Advanced Study in the Behavioral Sciences at Stanford, California.

Bell was cofounder and coeditor of *The Public Interest* and is now chairman of its publications committee. He has been on the editorial board of *The American Scholar* and now serves as a member of the Board of Editors of *Daedalus*.

During the Johnson administration, Mr. Bell was a member of the President's Commission on Technology, Automation and Economic Progress (1964–66), and served as cochairman of the Panel on Social Indicators of the Department of Health, Education, and Welfare (1966–68).

History, Technology and the Pursuit of Happiness

Edmundo O'Gorman

INTRODUCTION

It is fitting that at the outset of my talk I should express my thanks to Professor Daniel J. Boorstin, the distinguished historian and director of The National Museum of History and Technology, for having invited me to take part in these lectures sponsored by the Doubleday Company to mark its seventy-fifth year in publishing and to honor the memory of its founder, Frank Nelson Doubleday. It is no less fitting, therefore, to congratulate this highly respected publishing house for the anniversary and to wish that it may enjoy for many years to come a long and productive life.

Within the general program of these lectures, designed to explore the impact of technology on various fields of human endeavor, I have been asked to examine the role it has played in history. The complexity of such a formidable theme will be apparent to everyone, and it should not be surprising that in the brief space of a single lecture many aspects will perforce be left untouched and many strings dangling. And also, inevitably, it will be impossible to justify fully certain notions which must run the risk of appearing dogmatic or even as arbitrary assertions. Such deficiencies and danger did not weigh enough, however, to deter me from the attempt, if and when I may rest, as I very much hope, on the benevolence of my audience. It hardly seems necessary to warn that the subject of our concern involves nothing less than a vision of universal history and that, therefore, a previous notion about man must be basic to our inquiry, a notion which envisages man as an entity projected towards the future, responsible for his own being and, as we shall see, inexorably engaged in technological pursuits. I gladly acknowledge my debt to the many eminent historians and philosophers beginning with Dilthey, —to only mention a modern—who have forged such a fundamental notion to apprehend historical reality and the meaning of its happening. But in this connection it would be unfair not to specially remember the late Spanish philosopher José Ortega y Gasset, whose essay *Meditation on Technique* (Revista de Occidente, Madrid, 1939) has been particularly valuable.

Temixco, the twenty-fourth of November of 1972

Edmundo O'Gorman

I

There was a time, recent in years but historically remote, when intellectuals and artists held everything connected to technology as alien to culture. A man of culture was not and did not wish to be concerned with that dreary world of engineers, industrialists, and factory hands and only took note of its existence as some inferior sphere, necessary perhaps, but severed and even hostile to that heavenly circle, the habitat of the beautiful people and exquisite minds devoted to things spiritual.

The spokesmen for that assumed infra-orb reacted, for their part, with a parallel scorn towards those given over to books and to the cultivation of the higher arts, holding them in ridicule as a group of parasites who, masquerading behind an impenetrable barricade of tastes forbidden to common mortals and of an esoteric prattle, contributed in no way to mankind's progress and welfare.

This conflict between an outworn romanticism and a vulgar materialism predominated—explainably—in the United States at the time when American society was overrun by the most powerful industrial expansion to which history bears witness. It became a handy subject for satiric caricature and comic strips to depict the constant friction in upstart marriages, in which the woman, self-appointed priestess of artistic refinement and gentility, benefited from but slighted the activities of her husband, who, while a bold and farsighted adventurer in business and industry, exhausted his own spiritual cravings in the delights of the poker table or the baseball field.

Today, with nuclear weapons and space odysseys, nothing is left of that ridiculous situation except a vestige, on the one hand, in the jejune snobbism of a few interior decorators and, on the other hand, in the campaign perorations of a certain breed of politicians. But it is important to underline the legacy, because not infrequently technical activities are still disassociated from cultural pursuits, as if they belonged to respectively independent historical provinces. No one doubts any more the vast importance of technology, and even its contribution to the arts is widely accepted. Yet, ultimately, technology is viewed with deep suspicion and is envisaged as something susceptible of being considered separately from the true needs of human welfare. As something, therefore, which has had the greatest influence on history, shaping its course and marking it with indelible features, yet as something which merely *happened* to man, that is to say, as not essential or constitutive to his being. Technical activity is thus conceived of as an historical *accident*, a happening which happened to happen but which might not have happened at all. This misunderstanding of the true nature of technology is clearly evident in the touching but vain hopes of those who, believing it to be a sort of extraneous telluric force, would flee its evils to some pastoral refuge in perfect harmony with nature. It is that, indeed, which invites so many of the young today to embrace a so-called primitive life which at the end of a year or so will have victimized them with disease, discord, and unbearable tedium. Who hasn't undergone the appalling torture of a truly primitive picnic? The Victorians knew better: they held their picnics on the lawn in the cool shade of bygone medieval grandeur and thought of them as fitting occasions to read Horace and sip champagne.

II

In vigorous opposition to the notion of technique as incidental to man's life, Oswald Spengler[1] offers a firm starting

[1] Oswald Spengler, *Der Mensch und die Technik* (1931).

point. Technique, he tells us, is universal to all forms of life. It may not be ascribed to any one given period or species. It is, in essence, the assertion of all living entities in and as opposed to the environment or, if you wish, to nature. The existence of any form of life implies technique. This fundamental notion destroys the all too general supposition that technique is only concerned with the manufacture and use of tools and machines and discovers it to be the means by which all living beings are maintained in the natural circumstances in which they find themselves or, to say it in Spengler's words, it consists in the tactics employed by life to realize itself. The end pursued by technology, therefore, is one and the same as that pursued by life.

Such an omnifarious notion includes the whole biological scale from the amoeba to man, considering that all living beings have a particular way of behaving in the struggle which is living. But, clearly, as there are disparate forms and hierarchies of life, there also are different ways and refinements in technology. In comparison to the most sophisticated techniques of botanical life—the way certain plants reproduce, for instance—animal technique is ever so much richer and diversified, and here one may distinguish between passive technique—the camouflage of the chameleon, the flight of the gazelle—and the strategy and craftiness of the big predatory animals. Both imply a different way of life; the one seeks invulnerability, the other takes dangerous risks. Spengler places man among the predatory animals, and from this he derives his idea of history as a process prompted by the will to conquer and believes that human technology is the answer to that will, fully developed only, however, in Faustian civilization, exclusive to nations of Germanic stock.

Undoubtedly Spengler's powerful and alarming vision contains some truth, yet not enough to make us follow in its wake: after all, man surely is more than just a rapacious beast. But before turning from the tracts of this erstwhile most renowned historian-philosopher we may take advantage of one of his more apposite distinctions. He points out that all animal technique, whether passive or aggressive, possesses a common trait which sets it back at a gigantic distance from human technology—namely, that it does not introduce significant changes in the environment and, in any case, there is a void of consciousness as to the tremendous meaning attached to such an extraordinary effect. The life of an animal is like a toy in nature's hands and it remains inwardly identical, or, as Spengler says, in the interior recess of the soul. It is because of this narrow limitation that we can say with truth that animals live by instinct, which, of course, does not preclude intelligence, audacity, purpose, and admirable accomplishments.

Such a profound dissimilarity between animal and man warrants the exclusive usage of the word technology to human activities, and from here on we shall thus employ it.

III

Following Spengler's suggestion, we have asserted that the introduction of change in the environment is the specific trait of technology. Let us take a close look at this peculiarity, guided by Ortega's insight and authority.[2]

Man is not self-sufficient. In order to live, he must gratify his needs. All other forms of life are subject to the same prodicament, but the very remarkable difference is that while other forms submit to the circumstances in which they find themselves—an acquiescence which makes animals not to feel, strictly speaking, their needs as such—man alone is conscious of nature's hostility to the fulfillment of his wishes and, therefore, feels the constraints of nature as something imposed on him, a sort of injustice done to him and against which he rebels. Whatever it is he does not find at hand, awakens a state of mind and becomes thus subjective in such a way that it is felt as something which he lacks. In a word, that lack is conceived *sub specie* of necessity. Like the animals, man

[2] José Ortega y Gasset, *Meditación de la técnica, supra*, I.

is not self-sufficient, but the colossal difference is that he is conscious of that tragic condition.

As we go along, the incalculable consequences of such a peculiarity will become apparent. Let us fix our attention for the moment, on the most immediate. Indeed, when it so happens that nature provides, both man and animal simply and directly gratify their needs. But when that is not the case—which, as just explained, is what generates the feeling of necessity—man is moved to a very different kind of activity: it no longer consists in merely satisfying his want, for he must previously procure the means to do so. Obviously, heating oneself, for instance, is far from the same thing as having to build a fire in order to attain that end. We can clearly see, then, that this other kind of prior activity—exclusive to man—involves inventing processes and the manipulation of things for the purpose of having within reach at all times the means to satisfy whatever may have been felt as a need.

We may say, then, that the activity displayed by man in response to his feeling of a need resolves itself by imposing on the environment certain conditions or improvements which nature lacks. Such activity really means, therefore, that in order to live and fulfill the possibilities of his being, *man amends or, better, reforms nature in such a way that he may satisfy his wants.* And this is what technology is all about.[3]

IV

Let us take a closer look at our definition of technical activity. The first thing about it is that, like any other activity,

[3] It is hardly necessary to say that when we speak about man's needs we are not limiting them to his organic or biological wants. On the contrary, the gratification of these is but the previous condition for the truly human necessities to appear, necessities completely superfluous from the biological point of view and meaningless to an animal. On luxury as a chief objective of technology, cf. Ortega y Gasset, op. cit., II; Spengler, op. cit., V. 10.

it may be considered from the objective or subjective point of view, and our immediate concern will be to study in their turn these two aspects of the question.

It is obvious that, in saying that man amends or reforms nature, we are objectively saying that he introduces changes of the most varied kind, such as, for instance, altering the natural distribution of wooded land by razing forests; creating innumerable man-made things that are not naturally produced; or by inventing, say, with what are only vibrations in the air, what we know as music. In other words, by means of his technical activity man manufactures a special and new environment, an artificial nature different from although supported by the original, pristine nature.

In the remote dawn of man's history this other nature was hardly visible, because few changes in the environment were needed to satisfy primitive biological needs and spiritual wants: a footpath in the forest, a small ditch, rudimentary tools and domestic objects, rupestrian inscriptions and clay or wooden images to appease the wrath of the mysterious and menacing forces which beset man on all sides. But in essence those modest marks of man's technical activity and ingenuity have the same meaning as the colossal transformation that modern technology has imposed and will further impose on the face of the earth. Each transformation, the great and the small, gives witness to man's constructive power in changing the natural environment so that he may live in it according to his needs or wishes, because, as we shall see, it comes to the same thing.

But of the greatest importance, we must emphasize, is the dramatic dualism implied in what we have explained. Indeed, man finds himself submerged in an environment which, though usually thought of as one unit, is really two coexisting entities, distinct both in extension and kind. To acquaint ourselves with their peculiar features it is advisable to identify them separately by their proper names: by *universe* we refer to unchanged natural reality, whereas we shall reserve the name of *world* to signify artificial, or man-made, nature.

The concept of the universe is inclusive, by definition, of all that exists. The universe is not man-made, and because of that, and in that sense, it does not belong to him. It belongs to God, whether conceived in fetishist, mythological, theological, metaphysical, or scientific terms, but always as the power responsible for reality.

On the other hand, the world is man-made, and because of that, and in that sense, it is his; it belongs to him, not to the Godhead, who has not and cannot have any use for it. The world is mankind's possession and tenure, the result of the imagination, labor, courage, and endurance of countless generations since the earliest age, at the time when man became man.[4]

Now, this duality is evident in two ways. The first refers to the extension, the second to the class or kind of the one and the other entities. Indeed, it is not difficult to see that the world, being finite, is lodged within the universe, like, let us say, a capsule surrounded on all sides by the infinite space of the universe; but also, the world being artificial, such a capsule is a sort of tumor in the immaculate body of the universe.[5] These two traits that traditionally have kept apart the idea of the universe from the idea of the world must be borne in mind, because we are to come back to them when, later on, we will try to explain the vast difference between contemporary scientific technology and that of earlier ages. the present, it will suffice to know that the reform introduced in nature by man does not merely mean the bringing about

[4] The use of universe and world as synonymous is equivocal. Later (note 12 *infra*,), we shall understand the reason for that usage. The true meaning of world is, however, preserved in common speech when, for instance, we say of a friend that his work and family are "his world," or when, in reference to the tastes and habits of a man in society we talk of him as a "man of the world."

[5] It is worthwhile noticing that in this twofold dichotomy we find the origin of that extraordinary sentiment that has accompanied man for centuries, namely, that he is a stranger in the universe and a rebel against its divinely ordered disposition—the sentiment of alienation conceived by religion as that of guilt derived from original sin.

of changes but, much more radically, a confrontation with another kind of reality, which—it may be surmised—will eventually end up by swallowing it entirely.

V

Having considered technical activity on its objective side, we are to look at it from the subjective point of view by inquiring after its consequences, if any, to man's being.

If, as has been explained, man rebels against adverse conditions of his environment and modifies it to suit his wants, even a cursory view of this action will show that such a lack of conformity does not really concern nature as such, but the kind of life allowed to man on her terms, which is not the same thing. The distinction will be clearer if we put it by saying that for man it is not enough to maintain himself in nature—as is the case with other living entities—but he must maintain himself in a better way than nature will allow. Man does not, purely and simply, want *to be* in nature; he wants to shape it; he wants and seeks his own *well-being.* Human life is not only a struggle for life; it is a struggle for a better life,[6] and precisely because that is what man aspires to, he is moved to change the environment and reform nature in accordance to his desire.

But here we must caution against the common misunderstanding of a better life as consisting of the possession of means in greater quantity or superior quality—or both. This is a fallacy. In the first place because there are many to whom a better life means exactly the opposite, but in the second place, and more fundamentally, because great number or superior quality of means do not merely change the quality of life; they change life itself by opening the possibility of a different life, or, as is usually and aptly said, of a "new life." The fallacy, therefore, lies in assuming that life is something previous to and beyond the act of living, and the absurd

[6] Cf. Ortega y Gasset, op. cit., III.

sequitur will be that life is not the living of it, but a series of happenings that concern it but do not in the least affect it. This, however, is not all, for it must be noted that the desire for a better life not only implies, as we have just seen, a different life but also the aspiration of being other than the way in which one is. Such a consequence is obvious, as will appear in a couple of instances. If I desire a different life, let us say an immortal life, it is really because I aspire *to be* immortal; if I wish to lead a gentle life, it is that I wish *to be* a gentle man. And here, suddenly, we have uncovered man's extraordinary and unique peculiarity: he is an entity capable of ceasing to be what he is at a given moment in order to be in a different way. We may thus understand that man's being is not—as tradition has it at least since Parmenides—an essence, that is to say, an unalterable, immutable substance like, for instance, that of a stone or a star. Man is not a *thing*; he is, according to Montaigne's dictum "undulating and mutable"; an entity where being is not static, but dynamic; an entity projected towards the future in an ever-changing process which, as far as we may know, only stops with death.[7]

We may now return to the chief point of our inquiry, and in the light of the foregoing explanations it will appear that man's lack of conformity with the kind of life that nature itself will grant him and the correlative wish for a better life are at bottom a lack of conformity with nothing less than with *what he is*. In that deep-rooted rebellion, therefore, we find the secret spring that incites and moves all technical activity, which thus discloses itself as the means in man's power to be what he desires to be. But then, the world—that artificial nature manufactured by man's technical activity—discloses itself, in turn, as the environment required

[7] Robert Jay Lifton, in his interesting book *Boundaries: Psychological Man in Revolution* (Vintage, New York, 1970), describes contemporary man as "Protean Man" in view of his extreme social mobility. But this seems to me singularly shallow. Man is "protean" not by historical accident but by constitution.

by man to become what he wishes to become; the only environment, therefore, where he may actually realize himself.

VI

But what is it that man wishes to be? Here, at last, is the crucial question. Up to this point we have only hinted at the answer when asserting that man wishes his well-being or that he desires a better life. Now, no one will object, I believe, that those assertions are different ways of saying the same thing—namely, that man aspires to be happy. We may then summarize our whole argument by saying that man's necessity to be happy is the fundamental impulse in human life and that this prompts all the other needs. It is the necessity of necessities.

This answer to our crucial question may be regarded as supremely unreliable, because few concepts—if any—are as relative or subjective as that of happiness. Furthermore, nothing is more unstable and fickle, as we all know from personal and painful experience. So with reason we may be told that our castle is built on moving sand.

There is, however, no cause for despondency, provided we remember that in answering our question we did not speak of *feeling* happy, but of *being* happy. The former refers to a contingent possibility, the latter to a permanent state. Let us take two examples to make our point. It will be granted, I assume, that the happiness felt by a woman in seeing the man she loves—but which, on his parting, will turn to misery—is not the same as the happiness of a monk who covets nothing the world can offer. Wherein lies the difference? Undoubtedly in the fact that the woman needs to see the man she loves, whereas the monk has no needs at all, and that is why we say about him, not that he *feels* happy, but that he *is* happy.

We may now understand that man's aspiration to be happy

is his desire to be exempt from want or, to say it more technically, to become a *non-necessitated entity.*[8]

This is, then, the decisive notion for a thorough comprehension of the ultimate objective of technology. Indeed, now we may see that the artificial reality which, technology imposes upon natural reality—the man-made world—is the attempt to bring about the requirements of an environment in which man may reach the condition of a non-necessitated entity: a sort of heaven on earth where the state of beatitude promised by religion after death will be possible during mortal life.[9] Another thing, of course, is whether man will succeed in such an enterprise or in the trying will destroy himself. We may have time to ponder this alternative after having considered technology, as we now understand it, within the process of historical events.

VII

If technical activity is moved by man's desire to be what he wishes to be (we now know what that is) this implies the capacity of being able to imagine previously the new life or way of being desired for the future. Lacking that capacity, the desire to be one way or another could not arise and neither would the impulse to reform nature. Man's life would be a repetitious process like that pertaining to other forms of life. In other words, man would not be human. Man's supreme faculty, consequently, is not reason; it is imagination, the truly "divine spark" that makes sense of the ancient mythological intuition of man having been made in God's likeness. Indeed, imagination conceives reality before it exists, a prodigious

[8] This obviously cannot mean that man will no longer *have* necessities. It means that he will no longer be in need, as explained in III, *supra.*

[9] All this explains the conception of God as, precisely, the non-necessitated entity per se, but it also reveals the goal of man's true ambition.

achievement, second only to the act of creating *ex nihilo*, reserved to the Godhead.

Logically prior to all technical activity, we find therefore an act of the imagination by virtue of which man visualizes in anticipation what it is he wants to be and conceives a program thereof. Such a program may well be called the "project of life," the proper understanding of which entails three points.

In the first place, it is not an abstraction. The project of life contains a concrete program, such as, for instance, those which have given birth to the ancient Eastern and New World civilizations, Greek-Roman antiquity, Christianity and Buddhism, and modern man of Western culture. This trait is all-important to grasp the function of a project of life as the guiding principle of technical activity, because it determines its kind and its concrete objectives. Some societies will erect pyramids while others will build temples in response to the way in which they conceive their gods. The wealth of forms and styles in art, the differences in institutions and social habits, and, in other words, all the variegated prospects offered on the stage of the great theater of universal history are but the visible forms of different ways in which man has imagined his life and destiny: the echoes of the diversity and profusion in the projects of life devised by man ever since he became worthy of that name.

In the second place, notwithstanding the variety of concrete historical projects of life, they all recognize a fundamental unity. The reason for this is that the supreme objective is always the same one, namely—as we already know—the attainment of happiness. Such a purpose, therefore, defines the project of life of all the historically possible concrete projects of life.

In the third place, the above being the case, one may ask which—if any—of the concrete historical projects that have been tried may be regarded as the most successful. Now, the answer to this question cannot be given unless we first determine the two requisites of what would be the ideal project of life imaginable.

A. The first requisite is that the happiness which is the objective pursued should be attainable in this life. That not being the case, the project in question would not be a project of life, but a project of death.[10]

B. The second requirement is that the project of life in question should prompt a change in the environment so that it may no longer oppose man's wishes for a better life. The corresponding technology of such a project would have to reform nature in such a way that (a) it would include all of it, and (b) that it would control and dominate it. Under those conditions, indeed, man would not only cease to suffer want, but the causes of such a possibility would have disappeared.

It may be surmised that, in describing the above conditions, we have had in mind the project of life of our modern Western Euro-American civilization, and it is now for us to show that it actually fulfills them.

VIII

As to the first requirement, a few words will suffice. It seems indisputable that our modern civilization is grounded on a program which envisions life here and now, and although Christianity postulates a world beyond the grave, such a belief has never been a serious obstacle to the pursuit of happiness in this life, but pre-eminently so since the Renaissance, when modern civilization has its true beginning. Pascal's famous argument is all too significant: it is a good gamble—so it runs—to believe in an afterlife, because if it does not exist

[10] Such is the case of the "true life" imagined by certain Eastern cultures where happiness may only be attained in afterlife on condition of dissolving individuality in the Great Universal Whole. The corresponding technology of such a project of life will not attempt to overcome natural obstacles and it will consist in passive acts such as contemplation, inactivity, and self-sacrifice and immolation.

there can be no loss, whereas if it exists, we may well lose all.

The second requisite calls for a much more detailed explanation. It will be remembered that it has two aspects, which we shall consider separately, the first being that the reform imposed by technology on nature must include the whole of it. Let us turn, then, to history in search of the moment when an event of such magnitude took place. We need not look long for it—it happened at the end of the fifteenth and the first decades of the sixteenth centuries and is misleadingly known as the discovery of America. Some ten years ago I attempted to disclose in a small book called *The Invention of America*[11] the profound spiritual revolution brought about by that most singular and memorable event, and here a condensed abstract must suffice.

The world at the time Columbus crossed the Ocean was conceived as an entity formed of three parts: Europe, Asia, and Africa. These were thought of as intrinsically different entities, and for physical and theological reasons it was impossible to admit the existence of other parts. The threefold division did not, therefore, constitute an arithmetical series; it was a closed hierarchical series. Consequently, the world's structure was definitively and permanently made, and nothing could change it. It occupied only a portion of the globe called the Island of the World, surrounded by the waters of the Ocean. The Ocean, as well as the rest of the terrestrial globe, belonged to cosmic space, a part therefore, of the universe, belonging to God. In this ancient conception, the world was a small province, lodged within the infinity of the universe, and in which God graciously allowed man to live, a sort of prison that did not even belong to man, since he had not made it. Man thus turned out to be a serf infinitely thankful for his prisonlike cosmic habitat.

Subsequent explorations made it experimentally clear that

[11] First published in Spanish (Fondo de Cultura Económica, México, 1958) and somewhat enlarged—published later in English (Indiana University Press, Bloomington, 1961).

the new-found lands could not be a part of Asia and, consequently, had no place in the threefold and closed structure of the world. The alarming result was that those lands had to be conceived as a "fourth part" of the world, and were given the name of America.

Now, the all-important point we want to make here is that the obliged acceptance of a fourth part of the world not only brought about the collapse of the old way of conceiving it, but that it prompted a new way in which the world suddenly overflowed its ancient barriers and embraced the whole terrestrial globe, as a consequence of the Ocean having lost its old status to become a mere geographical accident. But this process did not stop there, because—*nota bene*—the existence of a fourth part of the world necessarily implied the real possibility of the existence of a fifth, a sixth, a seventh part, and so on *ad infinitum*. In other words, the ancient closed, three-part, hierarchical series became an arithmetical series, that is, an unlimited series of parts of identical nature. The world, therefore, really ceased to be conceived as made of "parts" and, consequently, embraced the whole universe—there being no longer any way to distinguish one part from another.[12]

Such, then, is the great cultural revolution brought about by the oceanic explorations carried on during the late fifteenth and early sixteenth centuries and which—in a very real sense—prepared the ground for that other, better-known spiritual adventure called the Copernican revolution. Thanks to the audacity and endurance of those explorers and to the courage of the men who challenged the validity of the time-honored threefold structure of the world, the duality—in one of its aspects—between the universe and the world (cf. IV, *supra*)

[12] That is the reason why the terms universe and world can be and are used as synonymous (cf. note 4, *supra*). In the light of what has just been explained, it will not appear outlandish to say that Columbus' voyage was truly a voyage into space, while contemporary special trips are voyages not out of the world. The truth of this apparent paradox is reflected in the otherwise alarming indifference towards those trips, except as admirable mechanical achievements.

was finally overcome as man broke loose from his ancient cosmic prison and serfdom.

IX

Having shown when and in what way the world extended its realm so as to include the whole universe, I must now show when and how Western man took possession of it and made it his own.

This stupendous feat was definitively achieved at a time contemporaneous or nearly—not casually, of course—with the revolution we have just described. Indeed, throughout the millenniums up to the Renaissance, technology employed direct and basically elementary means, so the reforms man imposed on nature left her, so to say, virginal and internally intact. On the whole, technical activities did not go beyond the extraction and exploitation of raw materials, the change of external features of the original landscape, and the subjugation of plants, animals, and man himself. Muscle was essentially the sole source of energy. Now, such methods—without disparagement for the many and extraordinary results achieved—entail, more than a true domination over nature, mere exploitation and utilization of natural resources and raw material. In pointing this out we are, of course, referring to the gigantic difference in regards to modern technology. Indeed, we are all aware that there was nothing like it before the discovery, during the fifteenth and sixteenth centuries, of physical-mathematical science, which opened the road, no longer to merely utilizing this or that natural resource or to solving such and such a technical problem, but to harness, for any task or purpose, nothing less than the incommensurate energy contained in matter, the energy, therefore, which sustains universal reality.

From that time on, the universe ceased to be a mysterious entity endowed with secret qualities—virtues and essences—governed by God through the inscrutable intentions of His providence. It became a vast system of foreseeable relations between phenomena, capable of being stated in mathematical language and controlled by mechanical means. I have referred, obviously, to the invention of the machine, a contrivance of—in principle—unlimited power, endurance, and productivity, like the universe itself, which, indeed, was from then on conceived as nothing more than a gigantic and perfect machine, the mover of all machines and, consequently, the machine of all possible imaginable machines.[13]

In this case, as in the previous one, the universe was also invaded by man's made world, this time not merely in extension but in its very core. Thus the ancient duality—in its second aspect—between artificial and original reality (cf. IV, *supra*) was overcome.

We may, then, state the obvious conclusion: namely that, as surmised, Western technology, alone in historical experience, complies with the requirements and conditions already mentioned (cf. VII, *supra*) as indispensable to establish and build a world in which man may actually reach his supreme calling by achieving happiness in this mortal, fleeting life.

But to this conclusion we must add that modern scientific technology is not only pre-eminent, but the only one capable of attaining such a level of excellence. Clearly man can go no further in that direction, as there is no other universe which he can swallow. And that the *ne plus ultra* has been reached may be confirmed by pondering the evident historical fact that Western man's project of life has been universally accepted as the only possible one with a future. In our day there are no other civilizations besides our own, since one can hardly count as such the poor relics of different ways of life

[13] It will be clear that the tremendous implication in this mechanical conception of the universe is that man, not God, is its true master, and here is disclosed the real and ultimate allurement in the project of life of Western culture. Little wonder, then, that authentic religious sentiment has always been wary of science as a manifestation of Satan's price and has linked technology to the diabolical arts.

which have survived but are inevitably doomed to disappear or live on only, if lucky, in the exhibit halls of the Smithsonian Institution. No one will contest, for instance, that Mao's China with its atomic bomb and its Western Marxist doctrine is anything else but a late and conspicuous daughter of a culture which out of habit we still call Western but should designate as universal. And where the extremity of the situation brought about by the uniqueness and excellence of scientific technology may be strikingly and dramatically evidenced is in that it has put humanity in the very real possibility of self-destruction. The enormity of such a circumstance should suffice to make us aware that since the inception of scientific know-how, history entered an intrinsically new phase in its perhaps tragic course.[14]

Without undue immodesty, I believe that by having shown what seems to me to be the true nature of all technical activity, and its fundamental purpose of establishing the necessary conditions for man to achieve happiness in this life, and by pointing out why scientific technology is the one and only way to reach that supreme goal, I have acquitted myself—to the best of my ability—of the task I undertook by accepting the distinction of partaking in this series of lectures. Here, therefore, I could put an end to this talk; it does seem appropriate, however, to say a few words concerning what appears to be the basic alternative for the future.

X

What we have really described is man's crowning victory over nature's obstacles to his happiness. Once this triumph is achieved, it follows that man's program for the future is to take advantage of his success and establish on this earth the promised paradise. This, however, is much easier said

[14] Quite obviously, the whole of our present inquiry implies an underlying conception of universal history, which would be interesting—or so we think—to describe at leisure or on another occasion.

than done, because—and here we have the real problem—the world built and sustained by technology comprises a tragic paradox. Indeed, if it is true that in such a world nature's obstacles are or may be abolished, it is none the less true that that same world begets, in its turn, new obstacles, of another kind, which, because of that, may not be overcome by the same means as the others. Man, consequently, is faced with a whole new set of pressures due to adverse conditions of his own making. And in this respect we must carefully avoid the snare of incriminating technology on that score. That amounts to fleeing from truth by the handy means of shunning the responsibility of whatever befalls.

Technology of itself is neither good nor evil, and to blame it is like reproaching the iceberg for having sunk the *Titanic*. Obviously, the sin is not to be found in technology but in the use to which it may be put. But here again we must be careful, because it is not a question—as it is frequently believed—of merely the reckless carelessness in the production and use of technical devices which has brought about ungovernable cities, pollution, and ecological unbalance. These and other, similar real tremendous evils, notwithstanding the gigantic task involved in curbing and remedying them, do not, however, involve the true problem. After all, they are basically technical problems to be technically solved, though it will undoubtedly mean deep change in social, political, and economical structures.

The great sin—let us call it that—is in having recourse to technology to attain ends alien to its native objective of making man's happiness a historical possibility by freeing him from want. To succumb to the temptations, for example, of using the great power of technological enterprise to achieve predominance at the cost of the well-being of the great masses in underdeveloped areas; to implement heinous doctrines involving the extermination of a whole people; or, more subtly, to transform a world so patiently and fearlessly built for man's well-being into its exact opposite—these are corruptions of man's historic imperative. In these examples we have instances of man's own high treason against his historic struggle to liberate himself from his original, animal condition. The propitious and beneficial artificial, man-made world is in jeopardy of becoming an ugly and brutish world in which man will be but a slave to his victory of freedom over nature, his most glorious historical achievement.

Clearly, such a peril now summons man to embark on a bold new enterprise comparable in design and dimension to the war he waged against nature's original opposition, and, consequently, to find the adequate technology to attain victory. We do not have time now to further dwell on this most vital question and may only indicate that such a technology must be addressed to the conquest of internal nature as the other was addressed against the hazards of external nature. Because if in this way man became the master of the universe, it is only too patent that he has yet to become the master of himself. Let us, then, put an end to our reflections by saying that man's project of life for the future, whatever else it might be, must include a program of self-restraint, education, and love, so that, in conquering what I do not hesitate to call spiritual innocence, he may regain paradise lost. The alternative—it goes without saying—is Big Brother or, perhaps mercifully, thermonuclear apocalypse.

EDMUNDO O'GORMAN

Mexican historian Edmundo O'Gorman teaches the philosophy of history and colonial Mexican history at the National University of Mexico, of which he is Professor Emeritus.

In 1958, in a series of lectures on "The Invention of America" (published by the Indiana University Press, Bloomington, Ind., 1961), O'Gorman joined philosophy and history in a subtle and original study of sixteenth-century Europe's bafflement at the knowledge of the existence of an unsuspected continent. Throughout a long and varied career, his special interest has been the history of the exploration and settlement of Spanish America.

Writer of numerous books, his most recent *La supervivencia política novo-hispana* (1969), O'Gorman also has edited the works of Joseph de Acosta (*Historia natural y moral de las Indias*, 1962), Bartolomé de Las Casas (*Apologética historia*, 1967), and Toribio Motolinío (*Memoriales*, 1971).

He has lectured at numerous institutions in the United States, and spent a one-year term as visiting professor at Brown University.

Technology and Evolution

Sir Peter Medawar

I

Genetic and exogenetic heredity

The use of tools has often been regarded as the defining characteristic of *Homo sapiens,* i.e., as a taxonomically distinctive characteristic of the species. But, in the light of abundant and increasing evidence that subhuman primates and even lower animals can use tools, the view is now gaining ground that what is characteristic of human beings is not so much the devising of tools as the communication from one human being to another of the know-how to make them. It was not so much the devising of a wheel that was distinctively human, we may suggest, as the communication to others particularly in the succeeding generation, of the know-how to make a wheel. This act of communication, however rudimentary it may have been even if it only took the form of a rudely explanatory gesture signifying "It's like this, see," accompanied by a rotatory motion of the arm—marks the beginning of technology, or of the science of engineering.

Everyone has observed with more or less wonderment that the tools and instruments devised by human beings undergo an evolution themselves that is strangely analogous to ordinary organic evolution, almost as if these artifacts propagated themselves as animals do. Aircraft began as birdlike objects but evolved into fishlike objects for much the same fluid-dynamic reasons as those which caused fish to evolve into fishlike objects. Bicycles have evolved and so have automobiles. Even toothbrushes have evolved, though not very much. I have never seen Thomas Jefferson's toothbrush, but I don't suppose it was very different from the one we use today; the Duke of Wellington's, which I *have* seen, certainly was not. To some Victorian thinkers, facts like these served simply to confirm them in the belief that evolution was the fundamental and universal modality of change. The assimilation of technological to ordinary organic evolution was not wholly without substance, because all instruments that serve us are functionally parts of ourselves. Some instruments, like spectrophotometers, microscopes, and radiotelescopes, are sensory accessories inasmuch as they enormously increase sensibility and the range and quality of the sensory input. Other instruments, like cutlery, hammers, guns, and automobiles, are accessories to our effector organs—not sensory but motor accessories.

A property that all these instruments have in common is that they make no functional sense except as external organs of our own: all sensory instruments report back at some stage or by some route through our ordinary senses. All motor instruments receive their instructions from ourselves.

It was for reasons like this that the great actuary and demographer Alfred J. Lotka invented the word "exosomatic," to refer to those instruments which, though not parts of the body, are nevertheless functionally integrated into ourselves. Everybody will have realized from personal experience how closely we are integrated psychologically with the instruments that serve us. When a car bumps into an obstacle, we wince more from an actual referral of pain than from a sudden premonition of the sour and skeptical face of an insurance assessor. When the car is running badly and labors up hills, we ourselves feel rather poorly, but we feel good when the car runs smoothly. Wilfrid Trotter, the surgeon, said that when a surgeon uses an instrument like a probe he actually refers the sense of touch to its tip. The probe has become an extension of his finger.

I do not think I need labor the point that this proxy evolution of human beings through exosomatic instruments has contributed more to our biological success than the conventional evolution of our own, or endosomatic, organs. But I do think it is worthwhile calling attention to some of the more striking differences between the two.

Genetic and exogenetic programs. By far the most important difference is that the instructions for making endosomatic parts of ourselves, like kidneys and hearts and lungs, are genetically programmed. Instructions for making exosomatic organs are transmitted through non-genetic channels. In human beings, exogenetic heredity—the transfer of information through non-genetic channels—has become more important for our biological success than anything programmed in DNA. Through the direct action of the environment, we do in a sense "learn" to develop a skin thicker on the soles of our feet than elsewhere. But information of this kind cannot be passed on genetically, and there is indeed no known mechanism by which it could be. It is only in exosomatic heredity that this kind of transfer can come about. We can learn to make and wear shoes and pass on this knowledge ready made to the next generation. Indeed, we even pass on the shoes themselves.

There is no learning process in ordinary genetic heredity: we can't teach DNA anything, and there is no known process by which the translation of the instructions it embodies can be reversed. No information that the organism receives in its lifetime can be imprinted upon the DNA, but in exogenetic heredity we can and do learn things in the course of life which are transmitted to the succeeding generation; thus exogenetic heredity is Lamarckian or instructional in style, rather than Darwinian, or selective. By no manner of means can the blacksmith transmit his brawny arms to his children, but there is nothing to stop his teaching his children his trade so that they grow up to be as strong and skillful as himself.

Learning as a new stratagem. The evolution of this learning process and the system of heredity that goes with it represents a fundamentally new biological strategem—more important than any that preceded it—and totally unlike any other transaction of the organism with its environment. In ordinary, endosomatic evolution and in cognate processes such as the so-called "training" of bacteria and, in immunology, antibody formation, we are dealing with what are essentially *selective*, as opposed to instructive, phenomena. The variants that are proffered for selection arise either by some random process such as mutation or by a process which it is not paradoxical to describe as a "programmed" randomness. By a "programmed randomness" I mean a state of affairs in which the generation of diversity is itself genetically provided for. Mendelian heredity provides for the preservation of genetic diversity for an unlimited period.

Exogenetic evolution reversible. Another important difference is this. Genetic evolution is conceivably reversible, just as it is thermodynamically conceivable that a kettle put on a lump of ice will boil. It's very unlikely, that's all. On the other hand, exosomatic evolution is quite easily reversible: everything that has been achieved by it can be lost or not reacquired. This is what specially frightens us when we contemplate the consequences of some particularly infamous tyranny that threatens to interrupt the cultural nexus between one generation and the next. This reversion to a cultural Stone Age is what each political party warns us will be the inevitable consequence of voting for the other. To bring the idea of reversibility to life, one should contemplate the plight of the human race if for any reason it did have to start again from scratch on a desert island: it is not heaven, but the old Stone Age, that lies about us in our infancy.

Popper's third world. I have been looking around in my mind for some one word or phrase to epitomize what I understand by our human inheritance through non-genetic channels —through indoctrination, that is, and the conscious transfer of information by word of mouth and through books. Karl Popper's[1] new book *Objective Knowledge* supplied the answer ready made. Let me therefore introduce you to Popper's concept of a "third world."

According to the philosophic views we specially associate with the name of George Berkeley, the apparently "real" world about us exists only through and by virtue of our apprehension of it. Thus sensible things and material objcets generally exist only as representations or conceptions or as "ideas" in the mind—hence the name "idealism." Berkeley argued persuasively, but Boswell very well knew that Berkeley's argument was of just the kind that would enrage Dr. Johnson. When Boswell teasingly said it was impossible to refute Berkeley's beliefs, "I refute it *thus,*" said Johnson, kicking a large stone so violently that he "rebounded" from it, thus simultaneously refuting Berkeley and confirming Newton's third-law of motion (the one about actions' having equal and opposite reactions).

However, even those who take a sturdily Johnsonian view of Berkeley's philosophy as it relates to the real world of material objects sometimes hold a Berkeleyan, or subjectivist, view of things of the mind. They tend to believe that thoughts exist by reason of being thought about, conceptions by virtue of being conceived, and theorems because they are the products of deductive reasoning.

Popper's new ontology[2] does away with subjectivism in the world of the mind. Human beings, he says, inhabit or interact with three quite distinct worlds: World 1 is the ordinary physical world, or world of physical states; World 2 is the

[1] Sir Karl Popper, the great philosopher and author of *The open Society and its Enemies.*

[2] Karl R. Popper. *Objective Knowledge—an Evolutionary Approach* (Clarendon Press, Oxford, 1972).

mental world, or world of mental states; the "third world" (you can see why he now prefers to call it World 3) is the world of actual or possible objects of thought—the world of concepts, ideas, theories, theorems, arguments, and explanations—the world, let us say, of all artifacts of the mind. The elements of this world interact with each other much like the ordinary objects of the material world: two theories interact and lead to the formulation of a third; Wagner's music influenced Strauss's and his in turn all music written since. Again, I mention for what it may be worth that we speak of things of the mind in a revealingly objective way: we "see" an argument, "grasp" an idea, and "handle" numbers, expertly or inexpertly as the case may be. The existence of World 3, inseparably bound up with human language, is the most distinctively human of all our possessions. This third world is not a fiction, Popper insists, but exists "in reality." It is a product of the human mind but yet in large measure autonomous.

This was the conception I had been looking for: the third world is the greater and more important part of human inheritance. Its handing on from generation to generation is what above all else distinguishes man from beast.

Popper has argued strongly that, although the third world is a human artifact, yet it has an independent objective existence of its own—and is indeed quite largely autonomous. I have already pointed out that the third world undergoes the kind of slow, secular change that is described as evolutionary,[3] i.e., is gradual, directional, and integrative in the sense that it builds anew upon whatever level may have been achieved beforehand. The continuity of the third world depends upon a non-genetical means of communication and the evolutionary change is generally Lamarckian in character, but there are certain obvious parallels between exosomatic evolution and ordinary, organic evolution in the Darwinian mode. Consider, for example, the evolution of aircraft and of auto-

[3] P. B. Medawar. *The Future of Man* (Methuen, London; Basic Books, New York, 1959).

mobiles. A new design is exposed to pretty heavy selection pressures through consumer preferences, "market forces," and the exigencies of function, by which I mean that the aircraft must stay aloft and the cars must go where they are directed. A successful new design sweeps through the entire population of aircraft and automobiles and becomes a prevailing type, much as jet aircraft have replaced aircraft propelled by airscrews.

I hope it is not necessary to say that the secular changes undergone by the third world do not exemplify and are not the product of the workings of great, impersonal historical or sociological forces. Just as the third world objectively speaking is a human artifact, so also are all the laws and regulations which govern its transformations. The idea that human beings are powerless in the grip of vast historical forces is in the very deepest sense of the word nonsensical. Fatalism is the most abject form of the aberration of thought which Popper calls "historicism." Its acceptance or rejection has not depended upon cool philosophic thought but rather upon matters of mood and of prevailing literary fashion. There was quite a fashion for fatalism in late-Victorian and Edwardian England, admirably exemplified by Omar Fitzgerald's famous stanza:

> 'Tis all a Checkerboard of Nights and Days
> Where Destiny with Men for Pieces plays
> Hither and thither moves, and mates and slays
> And one by one back in the Closet lays.

This is a comfortable doctrine, in so far as it spares us any exertion of thinking, but we may well wonder why it was so prevalent in late-Victorian and Edwardian England. The answer surely is that it fits very well with that high-Tory and latterly Fascist philosophy according to which, regardless of his upbringing, a man's breeding and genetic provenance fix absolutely his capabilities, his destiny and his deserts: a man not born a gentleman or, e.g., a German, could only at best merely simulate gentility or Germanity.

This kind of fatalism sounds very dated today, but we should ask ourselves very seriously whether there is not a tendency today to take the almost equally discreditable view that the environment has already deteriorated beyond anything we can do to remedy it—that man has now to be punished for his abandonment of that nature which, according to the scenario of a popular Arcadian daydream, should provide for all our reasonable requirements and find a remedy for all our misfortunes. It is this daydream that lies at the root of today's rancorous criticism of science and the technologies[4] by people believe, and seem almost to hope, that our environment is deteriorating to a level below which it cannot readily support human life. My own view is that these fears are greatly and unreasonably exaggerated.[5] Our present dilemma has something in common with those logical paradoxes that have played such an important part in mathematical logic. Science and technology are responsible for our present predicament but offer the only means of escaping the misfortunes for which they are responsible.

The coming of technology and the new style of human evolution it made possible was an epoch in biological history as important as the evolution of man himself. We are now on the verge of a third episode, as important as either of these: that in which the whole human ambience—the human house—is of our own making and becomes as we intend it should be: a product of human thought—of deep and anxious thought, let us hope, and of forethought rather than afterthought. Such a union of the first and third worlds of Popper's ontology is entirely within our capabilities, provided it is henceforward made a focal point of creative thought.

The word "ecology" has its root in the Greek word "oikos," meaning "house" or "home." Our future success depends

[4] P. B. Medawar, and J. S. Medawar, in *Civilisation and Science; in Conflict or Collaboration?* Ciba Foundation Symposium (North-Holland, London and New York, 1972).

[5] P. B. Medawar. *The Hope of Progress* (Methuen, London, 1972; Doubleday, New York, 1974).

Technology and the Limits of Knowledge

Arthur C. Clarke

Technology and Knowledge

The twin subjects of this talk are *technology* and *knowledge*, and whenever I hear that second word I am reminded of a little poem popular at Oxford about a hundred years ago:

> I am the Master of this College;
> What *I* don't know isn't knowledge.

This claim was, of course, immediately challenged by a rival establishment:

> In all Infinity
> There is no-one so wise
> As the Master of Trinity.

Unless my memory is betraying me yet again, the modest first couplet emanated from Balliol and was attached to Benjamin Jowett, the theologian and Greek scholar.

Today, of course, a man like Dr. Jowett lies squarely on the far side of the famous Culture Gap. Most of today's knowledge consists of things that he didn't know, and couldn't possibly have known. This is not because of the sheer increase in knowledge, though that has been enormous. But the very center of gravity of scholarship has now moved so far that there are vast areas where any high school dropout is better informed than the most highly educated man of a hundred years ago.

Much of this change may be linked with the other gentleman I mentioned just now: the Master of Trinity, than whom, et cetera. The most famous holder of this post was J. J. Thomson, discoverer of the electron; and *that* discovery marks the great divide between our age and all ages that have gone before. It transformed technology, and it transformed knowledge.

The electronic revolution, and the devices it has spawned, is now changing the face of our world and will determine the very structure of future society. And the discovery of the electron led, of course, directly to modern physics and the picture of the universe we have today—so much more complex and fantastic than could possibly have been imagined by any philosopher of the past.

One is almost tempted to argue that most *real* knowledge is a by-product of technology, but of course this is an exaggeration. Much that we know about the world around us has been derived over the centuries by simple naked-eye observation: in some important fields, like botany and zoology, this is still partly true. Yet, even here, we could never have understood the facts of simple observation without the technology represented by the microscope and the chemical laboratory. It can be argued that we do not really know anything until we understand it; mere description is not enough. Ancient naturalists such as Aristotle and Pliny recorded many of the basic facts of genetics; it is only in our time that the secret of the DNA molecule was uncovered, after a gigantic research effort involving every weapon in the technological armory from computers to electron microscopes.

There are those who think that this is a pity, and who somehow feel that knowledge is "purer" in direct proportion to its lack of contamination with technology. This—literally! —mandarin attitude is a consequence, as J. D. Bernal has remarked, of "the breach between aristocratic theory and plebeian practice which had been opened with the beginning of class society in early civilization and which had limited the great intellectual capacity of the Greeks."[1]

This failure of the Greeks—and the Chinese—to fuse technology and knowledge in a truly creative manner is one of the great tragedies of human history; it lost us at least a thousand years. Both these great civilizations had plenty of technology, some of a very high order, as Joseph Needham and others have shown. Nor did the Greeks despise it, as is often imagined; the myth of Daedalus and the reality of Archimedes show their regard for sophisticated mechanics.

Yet, somehow, these brilliant minds—of whom it has also been truly said that they invented all known forms of government and couldn't make one of them work—missed the breakthrough into experimental science; that had to wait for Galileo, two thousand years later. How near the Greeks came to the modern age you can see, if you have sufficient influence and persistence—it took me three visits and a letter from an admiral—in the basement of the National Museum of Athens. For there, tucked away in a cigar box, is one of the most astonishing archaeological discoveries of all time, the fragments of the astronomical computer found by sponge divers off the island of Antikythera in 1901. To quote from Dr. Derek Price: "Consisting of a box with dials on the outside and a very complex assembly of gear wheels mounted within, it must have resembled a well-made *18th Century* (my italics) clock. . . . At least twenty gear wheels of the mechanism have been preserved, including a very sophisticated assembly of gears that were mounted eccentrically on a turntable and probably functioned as a sort of epicyclic or differential gear system."[2]

Looking at this extraordinary relic is a most disturbing ex-

[1] *Science in History*, Vol., pp. 375–76 (Pelican Edition, London, 1969).

[2] "An Ancient Greek Computer," *Scientific American*, June 1959. On the copy he sent to me, Derek Price has written hopefully, "Please find some more." I am afraid that the most advanced underwater artifact I have yet discovered is an early-nineteenth-century soda-water bottle.

perience. Few activities are more futile than the "What if . . ." type of speculation, yet the Antikythera mechanism positively compels such thinking. Though it is over two thousand years old, it represents a level which our technology did not reach until the eighteenth century.

Unfortunately, this complex device merely described the planets' apparent movements; it did not help to *explain* them. With the far simpler tools of inclined planes, swinging pendulums, and falling weights, Galileo pointed the way to that understanding, and to the modern world.

If the insight of the Greeks had matched their ingenuity, the industrial revolution might have begun a thousand years before Columbus. By this time we would not merely be pottering around on the moon; we would have reached the nearer stars.

One of the factors which has caused this gross mismatch between ability and achievement is what might be called intellectual cowardice. In the extreme case, this is best summed up by that beloved cliché from the old-time monster movies: "This knowledge was not meant for man." Cut to the horrified faces of the villagers, as the mad scientist's laboratory goes up in flames. . . .

The non-celluloid version is a little less dramatic. It consists of assertions that something can never be known, or done, rather than that it *shouldn't* be. But often, I think, the underlying impulse is fear, even if the only danger is the demolition of a beloved theory. Let me give some examples which are relevant to the theme of this talk.

It's grossly unfair to judge anyone by a single piece of folly; few of us would survive such a critique. But I have never taken Hegel seriously—and have thus saved myself a great deal of trouble—because of the *Dissertation on the Orbits of the Planets,* which he published in 1801.

In this essay, he attacked the project then under way to discover a new planet occupying the curious gap between Mars and Jupiter. It was philosophically impossible, he explained, for such a planet to exist. . . . By a delightful irony of fate, the first of the asteroids had *already* been discovered a few months before Hegel's unfortunate essay appeared. I do not know if he issued a revised edition, but Gauss remarked sarcastically that this paper, though insanity, was pure wisdom compared to those that Hegel wrote later. . . .[3]

Some of my best friends are Germans, but I cannot resist quoting an even more splendid specimen of Teutonic myopia. When Daguerre announced his photographic process in 1839, it created such a sensation that some people simply refused to believe it. A Leipzig paper "found that Daguerre's claims affronted both German science and God, in that order: "The wish to capture evanescent reflections is not only impossible, as has been shown by thorough German investigation, but . . . the will to do so is blasphemy. God created man in his own image, and no man-made machine may fix the image of God. . . . One can straightway call the Frenchman Daguerre, who boasts of such unheard-of things, the fool of fools."[4]

One would like to know more about the subsequent career of this critic, whom I have resurrected not merely for comic relief but because he provides an excellent introduction to a much more instructive *débâcle*. This time, I am happy to say, the culprit is a Frenchman; I would not like anyone to accuse me of nationalistic bias. He is the philosopher Auguste Comte.

In the second book of his *Course of Positive Philosophy* (1835) Comte defined once and for all the limits of astronomical knowledge. This is what he said about the heavenly bodies:

> We see how we may determine their forms, their distances, their bulk, their motions, but we can never know anything of their chemical or mineralogical structure; and much less, that of organized beings living on their surface. We must keep carefully apart the idea of the solar system and that of the

[3] Willy Ley. *Watchers of the Skies* (Viking, New York, 1963), p. 320.

[4] *Light and Film* (Time/Life Books, New York, 1970), p. 50.

> universe, and be always assured that our only true interest is with the former. . . . The stars serve us scientifically only as providing positions. . . .

Elsewhere, Comte pointed out another "obvious" impossibility; we could never discover the temperatures of the heavenly bodies. He thus ruled out even the *theoretical* existence of a science of astrophysics; it is therefore doubly ironic that within half a century of his death, most of astronomy was astrophysics, and scarcely anyone was concerned with the solar system, which he claimed was "our only true interest."

The demolition of Monsieur Comte was produced by a single technological development: spectroscopy. I don't see how we can blame Comte for not imagining the spectroscope; who could possibly have dreamed that a glass prism would have revealed the chemical composition, temperature, magnetic characteristics, and much else, of the most distant stars? And can we be sure that, even now, we have discovered all the ways of extracting information from a beam of light?

Let us think about light for a moment, as the development of optics provides the most perfect example of the way in which technology can expand the frontiers of knowledge. Vision is our only long-range sense—unless one accept ESP—and until this century everything that we knew about the universe was brought to us on waves of light. If you doubt this, just close your eyes—or reread H. G. Wells' most famous short story:

> Núñez found himself trying to explain the great world out of which he had fallen, and the sky and mountains and sight and such-like marvels, to these elders who sat in darkness in the Country of the Blind. And they would believe and understand nothing whatever he told them. . . . For fourteen generations these people had been blind and cut off from all the seeing world; the names of all things of sight had faded and changed . . . and they had ceased to concern themselves with anything beyond the rocky slopes above their circling

> wall. Blind men of genius had arisen among them and questioned the shreds of belief and tradition they had brought with them from their seeing days, and had dismissed all these things as idle fancies, and replaced them with newer and saner explanations. Much of their imagination had shrivelled with their eyes. . . .

With the genius of the poet he pretended he wasn't, Wells created in this story a universal myth ". . . *their imagination had shrivelled with their eyes.*" And, on the contrary, how ours has expanded, not only with our eyes, but even more with the instruments we have applied to them!

Galileo and the telescope is the classic example; who has not envied him, for his first glimpses of the mountains of the moon, the satellites of Jupiter, the phases of Venus, the banked starclouds of the Milky Way? During those few months in 1609–10 there occurred the greatest expansion of man's mental horizons that has ever occurred in the whole history of science. The tiny, closed cosmos of the medieval world lay in ruins, its crystalline spheres shattered like the fragments of some discarded nursery toy. Which, in a sense, it was, being one of the childish things our species had to put aside before it could face reality. There will be other sacrifices to come.

One of the most remarkable things about the technology of handling light is the extreme simplicity of the means involved, compared with the far-reaching consequences. To give a humble but revolutionary example, consider spectacles.

The scholars of the ancient world, struggling to read by oil lamps and candles, must often have been functionally blind by middle age—especially as the manuscripts they studied were usually designed for art rather than legibility. The invention of eyeglasses (circa 1350) may well have doubled the intellectual capacity of the human race, for with their aid a man need no longer give up his work just when he was entering his most productive years. I don't know if this is an original idea—or whether it has already been refuted—but one could make a case for spectacles being a prime cause of the Renaissance.[5]

It is hard to think of a simpler piece of apparatus than a lens; yet what wonders it can reveal! Few people realize that the remarkable Dutch observer Van Leeuwenhoek discovered bacteria *with an instrument consisting of a single lens!* His "microscopes" were nothing more than beads of glass mounted in metal plates, yet they opened up a whole universe. Unfortunately, Van Leeuwenhoek was such a genius that no one else was ever able to match his skill, and the microscope remained little more than a toy; for its full use, it had to wait until Pasteur, two hundred years later.

And here is another of the great ifs of technology. Suppose Van Leeuwenhoek's observations had been followed up; then the germ theory of disease—often suggested but never proved—might have been established in the seventeenth century instead of the nineteenth. Hundreds of millions of lives would have been saved—and by this time the population explosion would have come and gone. Human civilization would by now have collapsed—or it would have safely passed through the crisis which still lies ahead of us—and for which, if you will excuse me, I have coined the word A*pop*aclypse.

The microscope and the telescope, both born about the same time, thus have sharply contrasting histories. The microscope remained a toy—the plaything of rich (well, fairly rich) amateurs like Samuel Pepys.[6] But the telescope, from

[5] As one of the undoubtedly countless examples of the need for eyeglasses in the ancient world, see Julius Caesar, Act V, scene II:

> *Cassius:* Go, Pindarus, get higher on that hill;
> My sight was ever thick; regard Titinius,
> And tell me what thou notest about the field.

And if anyone wants to know what Romans would have looked like wearing horn-rimmed spectacles, Phil Silvers has already obliged in *Something Funny Happened on the Way to the Forum.*

[6] On July 26, 1663, Pepys bought a microscope for the "great price" of five pounds, ten shillings. "A most curious bauble it is," which he used "with great pleasure, but with great difficulty."

Let us also never forget that, as President of the Royal Society,

the moment it was introduced, started a revolution in astronomy that has continued to this day.

Many years ago, however, the telescope came up against an apparently fundamental limit—that of *practical* magnifying power. Because of the wave nature of light, there is no point in using very high magnifications; the image simply breaks up, like an overenlarged newspaper block. Still, this natural limit is a very generous one. In theory, the Mount Palomar 200-inch would permit an incredible 20,000 power, which would bring the moon to within ten miles.

Alas, this delightful fantasy is frustrated by the medium through which the light must pass—the few dozen miles above the observatory. A star image can travel intact for a million million million miles, only to be hopelessly scrambled during the last microseconds of its journey by turbulence in the earth's atmosphere.[7] To the optical astronomer, all too often, the medium is the mess.

Even under the rare conditions of virtually perfect seeing, at mountaintop observatories, the highest magnification that can ever be used is only about 1,000. This means that under favorable conditions the smallest object that can be seen on the moon is about half a mile across, and on Mars about fifty miles across. But these figures are very misleading, because contrast plays a vital role. Lunar contrasts can be very high, owing to the starkness of the shadows. Mars contrasts are very low, making its surface features difficult to see and still harder to draw and photograph.

This tantalizing state of affairs led to one of the most famous, entertaining, and perhaps tragic episodes in the history of astronomy; I refer to the long controversy over the Martian

---

Pepys's name appears immediately below Newton's on the title page of the *Principia*.

[7] Very recently, it has been possible to approach the theoretical limits of magnification by the technique known as "speckle interferometry." The 200-inch Hale telescope has been used at an effective focal length of *half a mile* to photograph the disks of giant stars like Antares.

canals, which has been finally settled only during the last few months. It is an example of what can happen when the desire for knowledge outruns the technology of the time.

Though he was not the first man to "observe" the canals, Percival Lowell was certainly the man who put them on the map—and I use that phrase with malice aforethought. Carl Sagan has, perhaps unkindly, referred to Lowell as "one of the worst draftsmen who ever sat down at the telescope"; I have preferred to call him "the man with the tessellated eyeballs."[8]

Whatever Lowell's deficiencies as an observer, there can be no doubt of his ability as a propagandist. In a series of persuasive books, from 1895 onwards, he almost singlehandedly laid the foundations of a myth which was gleefully elaborated upon by several generations of science-fiction writers—of whom the most celebrated were Wells, Burroughs, and Bradbury. The ancient sea beds, the vast irrigation system which still brought life to a dying planet, the ruins of cities that would make Troy seem a creation of yesterday—it was a beautiful dream while it lasted, which was until July 15, 1965.

On that day, the overstrained technology of the telescope was surpassed—though by no means superseded—by that of the TV-carrying space probe. Mariner 4 gave us our first glimpse of the real Mars, though by another delightful irony of fate those initial pictures were almost as misleading as Lowell's fantasies. Not until Mariner 9's superb mapping of the entire planet, in 1972, did Mars slowly begin to emerge from the mists as an unique geological entity—and one of our main orders of business in the next hundred years.

In the other direction, down towards the atom, we have also broken through one apparently insuperable barrier after another. The optical telescope and the optical microscope reached their limits at about the same time, since these are both set by the wave nature of light. The wholly unexpected

[8] Both libels will be found in *Mars and the Mind of Man* (Harper & Row, 1973).

invention of the electron microscope suddenly increased magnifying power a thousandfold, allowing us to view structure of molecular size and producing advances in the understanding of living matter that could have been obtained in no other way.

In the last few years, there has been another breakthrough —I hate having to use this exhausted word, but there are times when there is no alternative. The Scanning Electron Microscope has done something quite new, and wholly beyond the power of the older optical and electronic instruments. By showing minute three-dimensional objects in sharp focus, it has allowed us for the first time to enter—emotionally, at least—the submicroscopic world. When you look at a good S.E.M. photo of some creature barely visible to the eye, you can easily believe that it is really as large as a dog—or even an elephant. There is no sense of scale; it is as if Alice in Wonderland's fantasy has come true and we are able to shrink ourselves down to insect size and have an eyeball-to-eyeball confrontation with a beetle.

The power of technology to change one's intellectual viewpoint is one of its greatest contributions not merely to knowledge but to something even more important: *understanding*. I cannot think of a better proof of this than some remarks made by Apollo 8's William Anders at the signing of the Intelsat Agreement here in Washington on 20 August 1971:

> Truly, the most amazing part of the flight was not the moon, but the view we had of the earth itself. We looked back from 240,000 miles to see a very small, round, beautiful, fragile-looking little ball floating in an immense black void of space. It was the only color in the universe—very fragile—very delicate indeed. Since this was Christmas time, it reminded me of a Christmas-tree ornament—colorful and fragile. Something that we needed to learn to handle with care. . . .

Now, the telescope, the microscope, even the rocket have given us only a change of scale or of viewpoint in *space*. In a more modest way, men have been achieving this ever since they started exploring the earth. What is very new in human history is the power to change our outlook on *time*.

The camera was the first breakthrough in this difficult area. From the beginning, the photographic plate could capture a moment out of time, in a manner never even conceived before—witness the quotation I have already given from that Leipzig newspaper. We all know the extraordinary emotional impact of old photographs; this is because they can provide, in a way not possible even to the greatest art, a window into the past. We can look into the eyes of Lincoln or Darwin; but not of Washington or Newton. At least, not yet. . . .

Because the first photographic emulsions were very slow, the earliest glimpses of the past were somewhat extended ones; they lasted minutes at a time. But about a hundred years ago the camera acquired sufficient speed to provide mankind with another wonderful tool; call it an image-freezer, or time-slicer.

It is in fact almost exactly a century ago (1872) that the flamboyant Eadweard (sic) Muybridge solved a famous problem that had baffled every artist since the creators of the first cave paintings. Does a running horse have all four feet off the ground at the same time? Muybridge found that the answer was "yes"; he also discovered that the characteristic "rocking-horse" position shown in innumerable paintings of charging cavalry and Derby winners was nonsense. This caused great heartburning in artistic circles.

Since then, of course, the camera has speeded up many millionfold. Until recently, the ultimate was a photograph which Dr. Harold Edgerton has hanging up in his office at M.I.T. It shows a steel tower surmounted by a globular cloud with three cables leading into it. The ends of the cables are a little fuzzy, which is not surprising. The cloud is an A-bomb, a few microseconds after zero. . . .

Yet now we have far surpassed that. Using laser techniques, a slug of light less than a centimeter long has been stopped in its tracks. I suppose we'll reach the end of the line when someone catches a single photon in mid-vibration. . . .

We are obviously a long way from the tempo of the running horse—just beyond the limits of human perception—or even of the hummingbird's wingbeat, which is still something that the mind can comprehend even if the eye cannot grasp it. Slicing time into thinner and thinner wafers has now led us into the weird world of nuclear phenomena, and perhaps even down to the atomic or granular structure to time itself. There may be "chronons," just as there are photons.

Now that every amateur photographer has a flash gun plugged into his camera, the power to freeze movement no longer seems such a miracle. And, of course, stopping time is not a *wholly* new experience to men, though in the past it was an uncontrollable one. A thunderstorm on a dark night was pre-twentieth-century man's equivalent of a modern strobe-light show, and must have impressed him even if he did not particularly enjoy it.

There may be worlds in existence which have natural strobe lighting, though until a few years ago not even the most irresponsible of science-fiction writers would have dared to imagine such a thing. For who could have dreamed of a star which switched itself on and off thirty times a second?

The Crab pulsar does just this, and it's strange to think that its flashes might have been discovered years ago—if anyone had been insane enough to look for them with suitable equipment. What would have happened to astronomy, I wonder, if that had been done? Many people would have been convinced that such a flickering star was artificial—and even now, I don't think we should dismiss this explanation. The pulsars may yet turn out to be beacons, and those who protest that they are a very inefficient way of broadcasting have been neatly answered by Dr. Frank Drake. How do we know, he has asked, that there aren't some *stupid* supercivilizations around?

Granted the improbability that it could have survived the initial supernova explosion, it's fascinating to speculate about conditions on a world circling the Crab pulsar. To our eyes, daylight would appear to be continuous, but it would really by 30 cycles per second A.C. A rapidly moving object would break up into discrete images. Something that appeared to be stationary might be really spinning at high speed.

What would be the effect of this on evolution? Could predators take advantage of these weird conditions to deceive their victims? One day I may work this idea up into a story. Meanwhile if Isaac Asimov (to take a name at random) uses it first, you'll know who he stole it from.

So far I have talked about slowing down time, but what about speeding it up? Of course, that's much easier; it requires very simple technology, but lots of patience. Though the results are often fascinating, I do not know if they have yet contributed much to scientific knowledge. Time-lapse films of clouds and growing plants are the best-known examples in this field. Anyone who has watched the fight to the death between two vines, striking at each other like serpents, will have a new insight on the botanical kingdom. And I hope that the meteorologists will learn a great deal from the global cloud-movement films that have been taken from satellites; eventually these may give us a synoptic view of the seasons and even long-term climatic changes.

This was anticipated by two great writers of the last century; first:

> . . . Night followed day like the flapping of a black wing . . . and I saw the sun hopping swiftly across the sky, leaping it every minute, and every minute marking a day. . . . The twinkling succession of darkness and light was excessively painful to the eye. Then, in the intermittent darkness, I saw the moon spinning swiftly through her quarters from new to full, and had a faint glimpse of the circling stars. Presently, as I went on, still gaining velocity, the palpitation of night and day merged into one continuous greyness. . . . the jerking sun became a streak of fire, a brilliant arch, in space; the moon, a fainter, fluctuating band. . . . Presently I noted that the sun belt swayed up and down, from solstice to solstice, in a minute or less, and that consequently my pace was over a year a minute; and minute by minute the white snow flashed

across the world, and vanished, and was followed by the bright, brief green of spring.

That, as I am sure you have all recognized, is from Wells' first—and greatest—novel, *The Time Machine.* Unlike his space romances, it is a book that could not have been written before the nineteenth century; only then had the geologists finally shattered the myth of Genesis 4004 B.C. and revealed the immense vistas of time that lie in the past—and may lie ahead. The emotional impact of that discovery on the more sensitive Victorians is preserved in Tennyson's famous lines[9]:

There rolls the deep where grew the tree.
O earth, what changes hast thou seen!
There where the long street roars hath been
The stillness of the central sea.

The hills are shadows, and they flow
From form to form, and nothing stands;
They melt like mist, the solid lands,
Like clouds they shape themselves and go.

We now know that this poetic vision is a pretty good description of continental drift, suddenly respectable after languishing for years in the wilderness somewhere to the southeast of Velikovsky. And what established continental drift was a series of breakthroughs in technology, quite as exciting as those involved in the exploration of space. We are accustomed to sending probes to the planets. We have now begun to send probes into the past.

I don't mean this literally, of course; I don't believe in time travel, though I understand that Kurt Gödel has shown that it is theoretically possible under certain peculiar and highly impracticable circumstances, involving the annihilation of most of the universe. The best argument against time travel, as has been frequently pointed out, is the notable absence

[9] *In Memoriam.*

of time travelers. However, a few years ago, one science-fiction writer pointed out a chillingly logical answer to this. Time travelers, like radio waves, may need a receiver . . . and none has been built yet. As soon as one is invented, we may expect visitors from the future . . . and we had better watch out.

Looking into the past, however, does not involve logical paradoxes, and our time probing has brought back knowledge which a few years ago would have been regarded as forever hidden. I wonder what Auguste Comte would have said if one had asked him the chances of finding the age of a random piece of bone, of locating the North Pole a million years ago, of measuring the temperature of the Jurassic ocean, or the length of the day soon after the birth of the moon? I feel quite certain that he would have said such things are as intrinsically unknowable as the composition of the stars. . . .

Yet such knowledge is now ours, and often through methods which, in principle at least, are surprisingly simple. Everyone is aware of the revolution in archaeology brought about by carbon 14 dating. The still more surprising science of paleothermometry depends on similar principles. If you measure the isotope ratios in the skeletons of marine creatures, you can deduce the temperatures of the seas in which they lived. So we can now go back along the cores taken from the ocean bed, and watch the rise and fall of the thermometer as the ice ages come and go, one after another. There is, surely, something almost magical about this. . . .

To track the wanderings of the earth's magnetic poles ages before compasses were invented or there were men to use them, appears equally magical. Yet once again, the trick seems simple—as soon as it is explained. When molten rock cools, which happens continually during volcanic eruptions, it becomes slightly magnetized in the direction of the prevailing field. The tiny atomic compasses become frozen in line, carrying a message which sensitive magnetometers can decipher.

But what about the length of the day millions of years ago? Surprisingly, this requires the simplest technology of all; merely a microscope, and infinite patience.

Just as the growth of a tree is recorded in successive rings, so it is with certain corals. But some of them show not only *annual* layering but much finer bands of *daily* growth. By studying these, it has been discovered that six hundred million years ago the earth spun much more swiftly on its axis; it had a 21-hour day, and there were 425 days in the year.

These remarkable achievements, and others like them, are allowing us to reconstruct the past like a gigantic jig-saw puzzle. Just how far can the process go? Is there *any* knowledge of the past which is forever beyond recovery?

A favorite science-fiction idea—though I have not seen it around recently—is the machine that can recapture images or sounds from the past. Many will consider that this is not science fiction, but fantasy. They may be right, but let us indulge in a little daydreaming.

There used to be a common superstition to the effect that no sound died completely, and that a sufficiently sensitive amplifier could recapture any words ever spoken by any man who has ever lived. How nice to be able to hear the Gettysburg Address, Will Shakespeare at the Globe, the Sermon on the Mount, the last words of Socrates. . . . But the naïve approach of brute-force amplification is of course nonsense; all that you would get is raw noise. Within a fraction of a second, all normal sounds, expanding away from their source at Mach One, become so dilute that their energy sinks below that of the randomly vibrating air molecules. Perhaps a thunder clap may survive for a minute, and the blast wave of Krakatoa for a few hours—but your words and mine last little longer than the breath that powers them. They are swiftly swallowed up in the chaos of thermal agitation which surrounds us; and when you amplify chaos, the result is merely more of the same commodity.

Nevertheless, there is a slight hope of recapturing sounds from the past—when they have been accidentally frozen by some natural or artificial process. This was pointed out a few years ago by Dr. Richard Woodbridge in a letter to the

I.E.E.E. with the intriguing title "Acoustic Recordings from Antiquity."[10]

Dr. Woodbridge first explored the surface of a clay pot with a simple phonograph pickup, and succeeded in detecting the sounds produced by a rather noisy potter's wheel. Then he played loud music to a canvas while it was being painted, and found that short snatches of melody could be identified after the paint had dried. The final step—achieved only after a "long and tedious search"—was to find a spoken word in an oil painting. To quote from Dr. Woodbridge's letter: "The word was 'blue' and was located in a blue paint stroke—as if the artist was talking to himself or to the subject."

This pioneering achievement certainly opens up some fascinating vistas. It is said that Leonardo employed a small orchestra to alleviate the Mona Lisa's boredom during the prolonged sittings. Well, we may be able to check this—if the authorities at the Louvre will allow someone to prowl over the canvas with a crystal pickup. . . .

A few months ago I wrote to Dick Woodbridge to find if there had been any further developments in this field—which, it should be pointed out, requires the very minimum of equipment; merely a pair of earphones, a phonograph pickup, a steady hand, and unlimited patience. But he had nothing new to report and ended his letter with a plaint which all pioneers will echo. "The bottom part of an S curve is a lonely place to be!" True, but there's a lot of room down there to maneuver. . . .

There must be better ways of recapturing sound, but I can't imagine what they are. Still less can I imagine any way of performing a much more difficult feat—recapturing *images* from the past. I would not say that it is impossible in principle; every time we use a telescope, we are, of course, looking backwards in time. But the detailed reconstruction of ancient images—paleoholography?—must depend upon technologies which have not yet been discovered, and it is probably futile to speculate about them at this stage of our ignorance.

It may well be hoped that—whatever the enormous bene-

[10] *Proceedings of the I.E.E.E.*, August 1969, pp. 1465–66.

fits to the historian—such powers never become available. There is a peculiar horror in the idea that, from some point in the future, our descendants may have the ability to watch everything that we ever do.

But, because a thing is appalling, it does not follow that it is impossible, as the H-bomb has amply demonstrated. The nature of time is still a mystery; there may yet be ways of seeing the past. Is that any stranger than observing the center of the sun—which is what we are now doing with telescopes buried a mile underground? Surely neutrino astronomy, involving the detection of ghostly particles which can race at the speed of light through a million million miles of lead without inconvenience—is a greater affront to common sense than a simple idea like observing the past.

I seem to be in grave danger, just when I am running out of time, of starting on an altogether new talk: "Knowledge and the Limits of Technology." Which only proves, of course, how difficult it is to separate the two subjects—or to establish limits to either.

In fact, no such limits may exist, this side of infinity and eternity. Those who fear that this is indeed true have often tried to call a halt to scientific research or industrial development; their voices have never been louder than they are at this moment.

"Well, there may be limits to growth, in the sense of physical productivity, though in a properly organized world we would still be nowhere near them. But the expansion of knowledge—of information—is the one type of growth that uses no irreplaceable resources, squanders no energy. In fact, in terms of energy, information provides some almost unbelievable bargains. The National Academy of Sciences' recent report on astronomy gave a statistic that I would not credit from a less reputable source. All the energy collected by our giant radiotelescopes during the three decades that have revolutionized astronomy is "little more than that released by the impact of a few snowflakes on the ground."[11]

In the long run, the gathering and handling of knowledge

[11] *Astronomy and Astrophysics for the 1970's* (NAS, 1973), p. 77

is the only growth industry—as it should be. And to make the enjoyment of that knowledge possible, technology must play its other great rôle: lifting the burden of mindless toil, and permitting what Norbert Wiener called "the *human* use of human beings."

We are only really alive when we are *aware*—when we are interacting with the universe at the highest emotional or intellectual level. Scientists and artists do this; so, to the limits of their ability, did primitive hunters, whose lives we are now completely reassessing in the light of new knowledge. Anthropologists have just discovered, to their considerable astonishment, that we lost the twenty-hour week somewhere back in the Neolithic. For in optimum conditions, a few hours of hunting/foraging a day were all that was needed to secure the necessities of life; the rest of the time could be spent sleeping, conversing, chewing the fat (literally)—and, of course, thinking.

But, as we have seen, thinking doesn't get you very far without technology. We can thank what I have christened the Agricultural-Industrial Complex for that. Unfortunately, this sinister organization also invented work and abolished leisure, which we are only now rediscovering, after a rather nasty ten thousand years. Hopefully, we will be able to make a safe transition into the postindustrial age, and then the slogan of all mankind will be—if I may change just two letters in Wilde's famous aphorism:

WORK IS THE CURSE OF THE THINKING CLASSES

It is technology, wisely used, that will give us time to think, and an unlimited supply of subjects to think about. And if it leads to our successors, either the intelligent computers or the "Giant Brains" that Olaf Stapledon described in his masterpiece *Last and First Men,* why should that be regarded as a greater tragedy than the passing of the Neanderthals?

Our technology, in the widest sense of the word, is what has made us human; those who attempt to deny this are denying their own humanity. This currently popular "treason of the intellectuals" is a disease of the affluent countries; the rest of the world cannot afford it.

Not long ago, I was driving through the outskirts of Bombay when I noticed a sadhu (holy man) with just two visible possessions. One was a skimpy loincloth; the other, slung round his neck on a strap, was a transistorized loud-hailer.

There, I told myself, goes a man who does not hesitate to use technology to spread his particular brand of knowledge. He has grasped the one tool he needs, and discarded all else.

And that is the true Wisdom—whether it comes from the East, or from the West.

ARTHUR C. CLARKE

Arthur C. Clarke, one of the best-known writers of science fiction and one of the most imaginative and best-informed explorers of the future, has written some forty-five books, more than thirty of them on space travel and space science. His first science fiction, *Prelude to Space* (1951), was preceded by two non-fiction books on space travel: an introduction to astronautics, *Interplanetary Flight* (1950), and *The Exploration of Space* (1951). His most recently published works are *The Lost Worlds of 2001* (1971), a short-story collection titled *A Wind from the Sun* (1972), and *Rendezvous with Rama* (1973).

Clarke collaborated with film producer Stanley Kubrick to make one of Clarke's short stories, "The Sentinel," into the movie *2001: A Space Odyssey*.

In 1945, Clarke proposed the concept of a communications satellite to relay global radio and television signals. The idea was dismissed at the time, but in 1965 the first such satellite was launched by COMSAT, actually using the orbit that had been plotted by Clarke.